D0849891

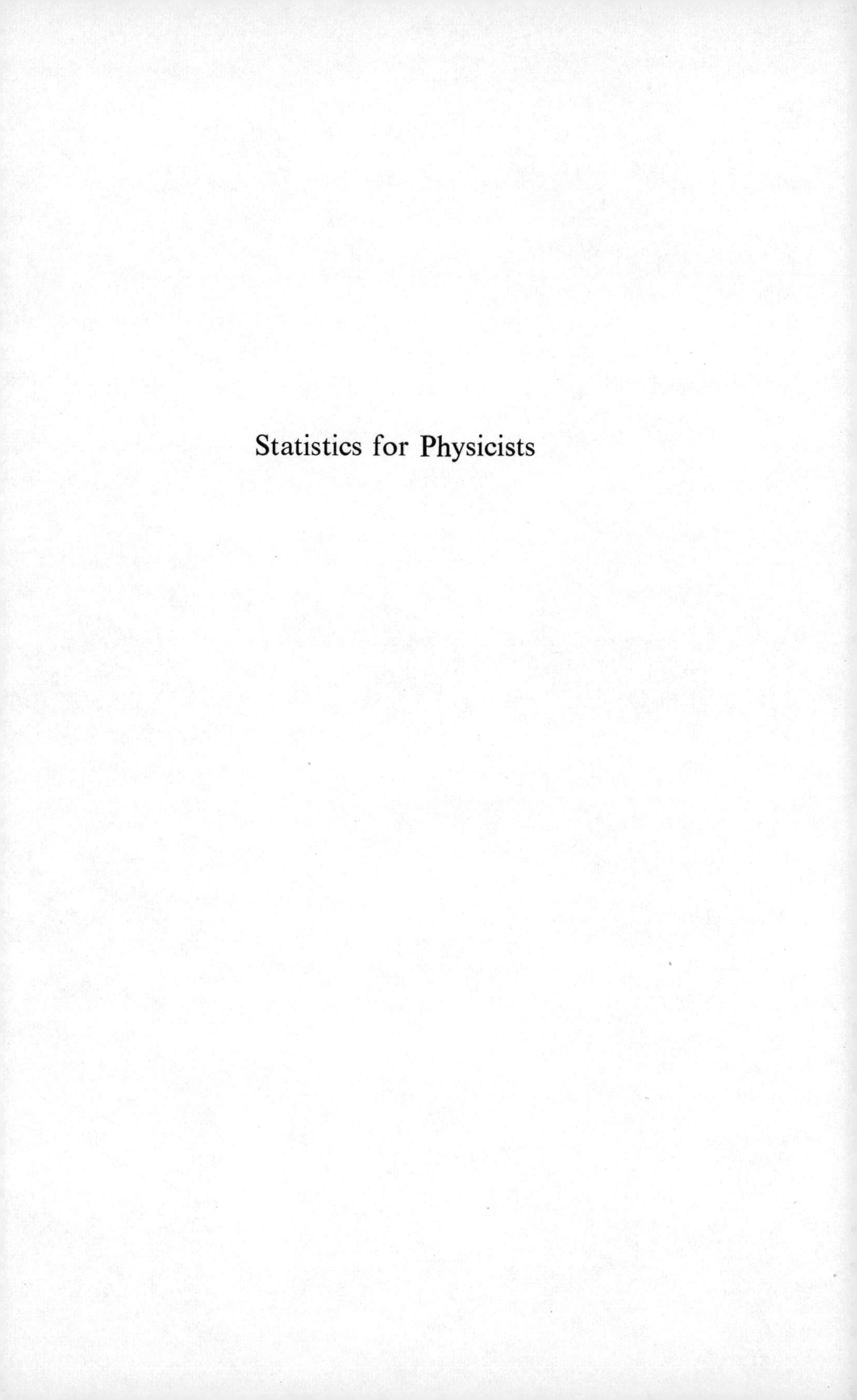

Statistics for Physicists

Statistics for Physicists

B. R. MARTIN
Department of Physics, University College London, England

1971

ACADEMIC PRESS · LONDON AND NEW YORK

ACADEMIC PRESS INC. (LONDON) LTD.
24–28 Oval Road,
London, NW1 7DD

U.S. Edition published by
ACADEMIC PRESS INC.
111 Fifth Avenue,
New York, New York 10003

ISBN: 0–12–474750–7
Library of Congress Catalog Card Number: 71–170764

Printed in Great Britain by
ROYSTAN PRINTERS LIMITED

To Darinka

Preface

This book has its origin in a set of notes on basic statistical ideas and techniques which I wrote in 1968 while a member of the Theoretical Physics Group at Brookhaven National Laboratory. In rewriting those notes, I have taken the opportunity to extend the discussion of some topics, and to include several new ones, with the result that the material in this book represents a far more detailed and systematic account of statistics as used by physicists. It should also be useful for other physical scientists and engineers.

I would like to express my gratitude to Professor Leonardo Castillejo, for generously reading most of the text, and for making many useful suggestions which have greatly helped to improve the presentation. Thanks are also due to Mrs. Janice Harring for patiently typing the manuscript.

Finally, acknowledgment is made to the McGraw–Hill Book Company, Cambridge University Press and the Biometrika Trustees, I.C.I. Ltd., and Butterworth and Co. Ltd., for their kind permission to reproduce the tables of Appendix D. I am also indebted to the Literary Executor of the late Sir Ronald A. Fisher, F.R.S., to Dr. Frank Yates, F.R.S., and to Oliver and Boyd, Edinburgh, for permission to reprint tables from their book "Statistical Tables for Biological, Agricultural, and Medical Research".

September, 1971 B. R. MARTIN

Acknowledgements

Acknowledgement is made to the following authors, editors and publishers whose material has been used in Appendix D, and for which permission has been received.

BIOMETRIKA TRUSTEES: E. S. PEARSON and H. O. HARTLEY,
Cambridge University Press
Biometrika Tables for Statisticians
Vol. 1 (1962) 130–131 Table D.5 χ^2 Distribution Function
Vol. 1 (1962) 157–163 Table D.7 F Distribution Function

THE CHEMICAL RUBBER CO.
CRC Standard Mathematical Tables, 14th ed., S. M. Selby, ed.
Table D.3 Binomial Distribution Function
Table D.4 Poisson Distribution Function

McGRAW–HILL BOOK COMPANY
Selected Techniques of Statistical Analysis, C. Eisenhart, M. W. Hastay, W. A. Wallis (1947) 284–309
Table D.10 Number of Observations for χ^2 Test of a Variance
Table D.11 Number of Observations for the Comparison of Two Variances using F-Test
Introduction to the Theory of Statistics, Mood and Graybill, 430–1.
Table D.1 Standard Normal Density Function
Table D.2 Standard Normal Distribution Function

OLIVER AND BOYD LTD., EDINBURGH
Design and Analysis of Industrial Experiments, O. L. Davies.
Research Vol. 1 (1948) 520–525
(1956) 606–607. Table D.8 Number of Observations for t-Test of Mean
(1956) 609–610. Table D.9 Number of Observations for t-Test of Difference Between Two Means
(1956) 613–614. Table D.10 Number of Observations for χ^2 Test of a Variance
Statistical Tables for Biological, Agricultural, and Medical Research, R. A. Fisher and F. Yates
(1938) 46 Table D.6 Student t-Distribution

Contents

1 Introduction

Although the applications of statistics in physics are numerous, many physicists' knowledge of the subject is uneven, albeit sometimes including a detailed knowledge of some practical technique. It was felt, therefore, that a short work which attempted to provide a brief, but fairly systematic, guide to the more commonly used statistical ideas and techniques, together with enough theoretical background to relate one idea to another, so that the whole does not just become a collection of "magic formulas", would be useful. This is the aim of these notes, and they are intended to be useful for the working physicist, both experimental and theoretical, as well as students at both graduate and undergraduate levels. However, it is not intended to imitate a class textbook, either in scope or rigour.

With respect to the scope; the book attempts to cover those areas which are more frequently used by physicists, rather than those invariably found in statistical textbooks. With respect to rigour; the author is a physicist, and although the proofs of theorems which are given will satisfy most other physicists they may not always be totally acceptable to the professional statistician.

Firstly, then, what do we mean by statistics? We may define "statistics" as that branch of scientific method which deals with data obtained by counting, or measuring, the properties of *populations*, where by "populations" we mean the collection of observations of a common attribute, e.g. the masses of a set of particles. Such populations are essentially numerical, and we will not normally consider the collection of particles themselves as a population. This is in accord with our everyday view of statistics, but it is interesting to note that this view has not always been held. For example, in 1770 statistics was defined in one book as, "The science that teaches us what is the political arrangement of all the modern states of the known world", and as late as 1834 we find, in the founding prospectus of the Royal Statistical Society, the following definition, "Statistics may be said to be the ascertaining and bringing together of those facts which are calculated to illustrate the conditions and prospects of society". However, despite such origins, statistics has rapidly become an extensive and highly developed branch of numerical science, concerned not just with data manipulation, but with such questions as the design of experiments, the principles of decision making, and many others, all of which have relevance to physics.

The plan of this book is as follows. To begin with, there is a very short introductory chapter on probability theory. A discussion of probability theory is an essential prerequisite before the subject of statistics can be meaningfully introduced, but no attempt has been made to give a detailed treatment of the topic, and the discussion instead relies largely on intuition. This brief chapter is followed by a discussion of theoretical distributions, Chapter 3 being concerned with the basic ideas and definitions, including definitions of those parameters which are used to characterize any finite population, and Chapter 4 describes the properties of several distributions frequently met in practice, together with others useful for illustrative purposes. Sampling theory is treated in the two chapters that follow. Chapter 5 is concerned with theoretical results, but at the end of this chapter the link is made between theoretical statistics and the experimental situation. Chapter 6 concentrates on the properties of three important sampling distributions associated with the normal distribution. These are the χ^2, F and Student t distributions. There follow three chapters on the important practical topic of point estimation, i.e. the estimation of the values of parameters. The first of these, Chapter 7, opens with a discussion of the general properties required of point estimators, and then describes the method of estimation known as Maximum Likelihood. Although the maximum likelihood method is, in a sense, the most general method of point estimation, nevertheless other methods are also widely used. One which is very popular is that of Least-Squares, and this is described in some detail in Chapter 8. The third of these three chapters, Chapter 9, gives a brief discussion of some other methods of point estimation. Point estimation has always, in practice, to be supplemented by a statement about the error associated with the estimate. This is the problem of interval estimation and is considered in Chapter 10, with particular reference to obtaining confidence intervals for the parameters of the important Normal distribution. Besides estimation the other main branch of statistics is that of hypothesis testing, and this is treated in one longer chapter, Chapter 11.

The eleven chapters are followed by a Bibliography and four Appendices, the latter being included in an attempt to make the book reasonably self-contained. The Bibliography is short and contains only those works which I have found particularly clear and useful. In Appendix A, Miscellaneous Mathematics, there are collected together various mathematical results and notations used in the text. Although all these topics have been met in the average physicist's undergraduate days nevertheless the reader may like to scan this Appendix first, to reacquaint himself with the necessary mathematics. Appendices B and C are concerned with some practical questions which arise in the course of estimation problems. In Chapter 8, on the Least-Squares method, the importance of using orthogonal polynomials is stressed, and an account of such functions is given in Appendix B. If the Least-Squares method is used in situations which are not linear, or if the Maximum Likelihood method of estimation is used, then the problem of minimizing, or maximizing, a function of several variables arises. Therefore, Appendix C contains a discussion of the problem of numerical optimization of a function of n variables. Finally, Appendix D contains a short set of statistical tables.

2 Probability Theory

2.1 Introduction

Statistics is intimately connected with that branch of pure mathematics known as probability theory, and before we can meaningfully discuss statistics we must say a little about probabilities. Probability theory as discussed in textbooks of mathematics uses such ideas as set theory and measure theory. We will use a minimum of such concepts but rely instead on more mundane methods (including the physicist's reputedly better intuition).

The term "probability" as used by mathematicians and physical scientists has two distinct meanings. For the former the parameters and the nature of the population are known, and can be specified exactly (e.g. by one of the mathematical forms to be discussed in Chapter 4). In this situation probability theory can be developed axiomatically. However, in physical situations the parameters of the population are rarely known (in fact one of the objects of statistical analysis is to obtain values for them), and the problem arises of determining which mathematical expression is correct when only a portion of the population is known. Without a knowledge of the entire population we cannot, of course, make absolutely precise statements concerning how the population is distributed, but we can make less precise statements in terms of a probability operationally defined as *the limit of the relative frequency of occurrence.*

Before proceeding to the definition of probability, however, we will need a few simple, but basic, definitions of some subsidiary quantities, which express more formally our intuition about such terms as "experiment", "event", etc.

Suppose we can set up a set of initial conditions that are reproducible. These conditions define an *experiment*, and by making an observation (or a set of observations) we produce an *outcome* of the experiment. The outcomes we will denote by x_i and they will be either single numbers, or possibly sets of numbers.

DEFINITION 2.1. The set of all possible outcomes x_i ($i = 1, 2, \ldots, n$) of an experiment is called the *sample space*, or *population*, and x_i is a *sample point* in the space.

DEFINITION 2.2. A subset of the sample points, e.g. $x_1, x_2, \ldots, x_m$ $(m < n)$ is called an *event* and will be denoted by

$$E \equiv \{x_i \mid i = 1, 2, \ldots, m\}.$$

If $m = n$, i.e. *all* the sample space is included in the event, then it will be denoted by

$$S \equiv \{x_i \mid i = 1, 2, \ldots, n\}.$$

The *occurrence* of an event may now be formally regarded as the situation where the sample point to which observations give rise is included in the subset of sample points defining the event.

For example, an experiment could be the tossing of a six-sided die, for which the outcomes would be one of the six numbers one to six. A simple event would also be one of these numbers, and the occurrence of the event would be the situation where the number defining the event was observed on the face of the die. A more general example of an experiment is the measurement of the heights of all males living within a given radius. Here the outcomes are not discrete, and an event would, in practice, be defined by a subset of sample points spanning a small interval of heights. Then the occurrence of the event would be interpreted as the situation where a measured height fell within the specified range.

We can now proceed to the operational definition of probability.

DEFINITION 2.3. Consider a sequence of n trials of an experiment in which the event E of a given class occurs n_E times. Then the ratio n_E/n is called the *relative frequency* of the event E, and is denoted $R[E]$. The *probability* $P[E]$ of the event E is the limit approached by $R[E]$ as n increases indefinitely, it being assumed that this limit exists.

Note that the above definition of probability differs somewhat from the mathematically similar one, i.e. that for some arbitrarily small quantity ε there exists a large n, say n_L, such that $|R[E] - P[E]| < \varepsilon$ for all $n > n_L$. The operational definition has an element of uncertainty in it derived from the fact that in practice only a finite number of trials can ever be made. This way of approaching probability is thus essentially experimental. One assigns an *a posteriori* probability to an event on the basis of experimental observation. The mathematical approach assigns an *a priori* probability to an event on the basis of a given mathematical model about the possible outcomes of the experiment. A typical situation that occurs in practice is when an experimentalist constructs a model of nature and computes from that model certain *a priori* probabilities concerning the outcomes of an experiment. The experiment is then performed, and on the basis of the results obtained certain *a posteriori* probabilities are calculated for the same events. The value of the model is then judged by the agreement between these two sets of probabilities, and on this basis modifications may be made to the model. These ideas will be put on a more quantitative basis in later chapters when we discuss estimation and the testing of hypotheses.

There are many other ways of defining probabilities, and statisticians do not agree amongst themselves on the best way. However, in this book we will not dwell too much on the differences between the various definitions.

2.2 Calculus of probabilities

As remarked above, there is no general agreement on the best way to approach the mathematical theory of probability, and so we shall pass straight to the rules of probability without discussing the controversial fundamentals. We shall start by listing a number of basic definitions which relate to our previous definitions.

DEFINITION 2.4. The probability of an event E is a number in the range 0 to 1, i.e.

$$0 \leqslant P[E] \leqslant 1, \quad \text{and} \quad P[E] = 1 \quad \text{if} \quad E \equiv S.$$

DEFINITION 2.5. Let E be an arbitrary event of an experiment. Then the event "not E" (the *complement* of E), will be denoted $\bar{E}$.

DEFINITION 2.6. The event "A or B", (the *intersection* of A and B), i.e. the event in which *both A and B* occur, is denoted $A \cap B$, or $B \cap A$. Thus if

$$A = \{x_i \mid i = 1, 2, ..., 10\},$$

and

$$B = \{x_i \mid i = 5, 6, ..., 20\},$$

then

$$A \cap B = \{x_i \mid 5, 6, ..., 10\}.$$

If $A \cap B = 0$ then the events are said to be *distinct*.

DEFINITION 2.7. The event "A or B" (the *union* of A and B), i.e. the event in which *either A or B or both* occurs is denoted $A \cup B$ or $B \cup A$. Thus using the example of Definition 2.6. we have

$$A \cup B = \{x_i \mid i = 1, 2, ..., 20\}.$$

These definitions can be given a simple geometrical interpretation by the diagram of Fig. 2.1.

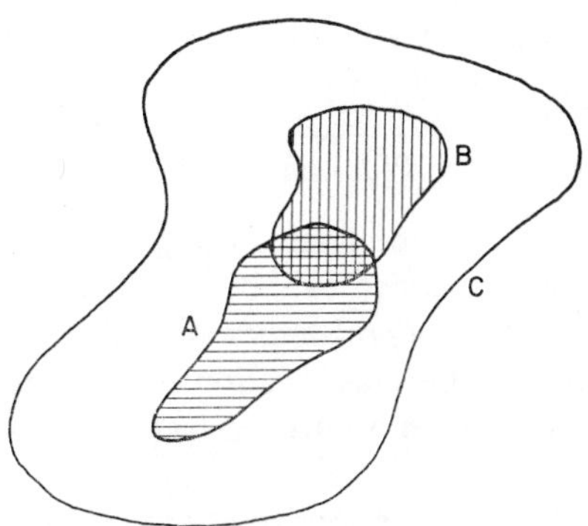

FIG. 2.1. The sample space S consists of all points within the boundary curve C. A and B are two events in S. The doubly-shaded area is $A \cap B$, and the sum of all the shaded areas is $A \cup B$. Finally, the unshaded area is $\overline{A \cup B}$.

DEFINITION 2.8. If an experiment can result in n mutually exclusive (i.e. the occurrence of one event precludes the occurrence of the others), and equally likely outcomes, n_B ($\neq 0$) of which correspond to the occurrence of event B, and n_{AB} of which correspond to the occurrence of the event A, given that B has occurred, then *the probability of event A given that B has occurred* is

$$P[A \mid B] = \frac{n_{AB}}{n_B}, \tag{2.1}$$

and is called *the conditional probability of A.*

If we use the operational form of probability, Definition 2.3, then Eqn (2.1) may be written

$$P[A \mid B] = \frac{P[A \cap B]}{P[B]}. \tag{2.2}$$

Note, however, that the price of not looking too closely at fundamentals has resulted in a definition which is somewhat circular, because of its use of the phrase "equally likely". A simple example will illustrate the use of Eqn (2.2).

EXAMPLE 2.1. Three "fair" coins are tossed, and we are told that at least two of them have fallen "tails". What is the probability that the third coin has fallen "heads"?

Let B be the event with at least two tails, and let A be the event with at least one head.

Then

$$P[A \cap B] = 3/8; \qquad P[B] = \tfrac{1}{2}.$$

Thus, from Eqn (2.2),

$$P[A \mid B] = 3/4,$$

i.e. the probability is 3/4 that the final coin has fallen "tails".

If the occurrence of an event can be classified according to multiple criteria then the term *marginal probability* is used whenever one or more of these criteria are ignored in the classification. If we consider the case of three classifications A, B and C, then we are led to the following definition.

DEFINITION 2.9. If the classifications under the criteria are $A_1, A_2, \ldots, A_r$; $B_1, B_2, \ldots, B_s$; and $C_1, C_2, \ldots, C_t$; with

$$\Sigma P[A_i] = \Sigma P[B_i] = \Sigma P[C_i] = 1,$$

then the *marginal probability* of A_i and C_k is

$$P[A_i \cap C_k] = \sum_{j=1}^{s} P[A_i \cap B_j \cap C_k], \tag{2.3}$$

and, likewise, the marginal probability of C_k is

$$P[C_k] = \sum_{i=1}^{r} \sum_{j=1}^{s} P[A_i \cap B_j \cap C_k]$$

$$= \sum_{i=1}^{r} P[A_i \cap C_k]$$

$$= \sum_{j=1}^{s} P[B_j \cap C_k].$$

The final definition concerns the concept of independence.

DEFINITION 2.10. The event A is *independent* of the event B if

$$P[A \mid B] = P[A]. \tag{2.4}$$

The above definitions may be used very simply to derive the following useful results:

$$P[\bar{A}] = 1 - P[A], \tag{2.5}$$

$$P[A \cap B] = P[A] \cdot P[B \mid A] = P[B] \cdot P[A \mid B]$$
$$= P[A] \cdot P[B] \text{ (if } A \text{ and } B \text{ are independent)}, \tag{2.6}$$
$$P[A \cup B] = P[A] + P[B] - P[A \cap B]$$
$$= P[A] + P[B] \text{ (if } A \text{ and } B \text{ are mutually exclusive)}. \tag{2.7}$$

Finally, we shall state the basic theorem of permutations:

THEOREM 1.1. *The number of ways of "permuting" (i.e. arranging) m objects selected from n distinct objects is*

$$nPm = \frac{n!}{(n-m)!}. \tag{2.8}$$

It follows from this theorem that the total number of "combinations" of the m objects without regard to arrangement is

$$nCm = \binom{n}{m} = \frac{nPm}{m!} = \frac{n!}{m!(n-m)!}. \tag{2.9}$$

2.3 Statistical inference

The calculus of probabilities as outlined above proceeds from the definition of probabilities of simple events to the probabilities of more complex events. In practice, however, what is required is just the inverse, i.e. given certain experimental observations we require to know something about the parent population and the generating mechanism by which they were produced. This, in general, is the problem of statistical inference. To illustrate a basic difficulty we shall discuss briefly a theorem propounded by the Rev. Thomas Bayes, an English clergyman, in 1763.

THEOREM 2.2. (Bayes' Theorem). *If B_i ($i = 1, \ldots, n$) are mutually exclusive and exhaustive (i.e. all possible events are included in the B_i) events, and if A can occur only in combination with one of the n events B_i, then*

$$P[B_i \mid A] = \frac{P[B_i]\, P[A \mid B_i]}{\sum_{j=1}^{n} P[B_j]\, P[A \mid B_j]}. \tag{2.10}$$

The proof of this theorem is very simple but affords an illustration of the use of Definitions (2.4) to (2.10).

Proof. From Eqn (2.6)

$$P[A \cap B_i] = P[A]\,P[B_i \mid A],$$

and

$$P[B_i \cap A] = P[B_i]\,P[A \mid B_i],$$

but these two quantities are equal and so

$$P[B_i \mid A] = \frac{P[B_i]\,P[A \mid B_i]}{P[A]}. \tag{2.11}$$

Now from Definition (2.9) and Eqn (2.6)

$$P[A] = \sum_{j=1}^{n} P[A \cap B_j]$$

$$= \sum_{j=1}^{n} P[B_j]\,P[A \mid B_j]. \tag{2.12}$$

Thus from (2.11) and (2.12)

$$P[B_i \mid A] = \frac{P[B_i]\,P[A \mid B_i]}{\sum_{j=1}^{n} P[B_j]\,P[A \mid B_j]}, \tag{2.13}$$

which completes the proof.

Bayes' Theorem has some important consequences and we shall examine it more closely. Suppose an event can be explained by the mutually exclusive hypotheses represented by B_1, B_2, ..., B_n. These hypotheses have certain *a priori* probabilities $P[B_i]$ of being true. Each of them can give rise to the occurrence of A, but with distinct probabilities $P[A \mid B_i]$. Bayes' Theorem tells us how to compute the *a posteriori* probabilities $P[B_i \mid A]$, which are the probabilities of having B_i *when A is known to have occurred.* The quantity $P[A \mid B_i]$ is called the *likelihood.* If we had to choose an hypothesis from the set B_i we would choose that one with the greatest *a posteriori* probability. However, Eqn (2.10) shows that this requires a knowledge of the *a priori* probabilities $P[B_i]$, and these are, in general, unknown. *Bayes' Postulate* is the hypothesis that, nothing known to the contrary, the *a priori* probabilities should all be taken as equal. We will give a simple example of the use of *Bayes' Postulate.*

EXAMPLE 2.2. Suppose we have a jar (or urn in statistical parlance) which contains four balls which may be either all white (hypothesis 1), or two white and two black (hypothesis 2). If n balls are drawn, one at a time, replacing them after each drawing, the probabilities of obtaining an event E with n white balls under the two hypotheses are

$$P[E \mid H_1] = 1; \qquad P[E \mid H_2] = 2^{-n}.$$

Now, from Bayes' Postulate

$$P[H_1] = P[H_2] = \tfrac{1}{2},$$

and so from Eqn (2.10)

$$P[H_1 \mid E] = \frac{2^n}{1 + 2^n}; \qquad P[H_2 \mid E] = \frac{1}{1 + 2^n}.$$

Thus, provided no black balls appear, we would always accept the first hypothesis because it has the larger *a posteriori* probability.

Bayes' Postulate has been the subject of much controversy. In the frequency theory of probability it would imply that events corresponding to the various B_i are distributed with equal frequency in some population from which the actual B_i has arisen. Many statisticians reject this as being unreasonable.

Later in this book we shall discuss some of the many suggested alternatives to Bayes' Postulate, including the principles of least-squares and minimum chi-square. We shall anticipate that discussion by mentioning briefly here (and in more detail in Chapter 7), one principle of general application, that of *maximum likelihood.*

From (2.10) we see that

$$P[B_i \mid A] \propto P[B_i]L, \tag{2.14}$$

where L stands for the likelihood. *The Principle of Maximum Likelihood* states, that when confronted with the choice of a set of hypotheses B_i, we choose that one which maximizes L, if one exists, i.e. that one which gives the greatest probability to the observed event. Note that this is *not* the same as choosing the hypothesis with the greatest probability. It is not at all self-evident why one should adopt this particular choice as a principle of statistical interference, and we will return to this point in Chapter 7. For the simple case in Example 2.2, the Maximum Likelihood method clearly gives the same result as Bayes' Postulate.

3 Theoretical Distributions: Basic Ideas

3.1 Population parameters

In most physical work the observer does not have a complete population, but has instead a *sample*, which is a subset of the total population. It is the central problem of physical statistics to estimate the properties of the population from the nature of the sample. This process is known as *statistical inference* and was briefly introduced in Chapter 2. Firstly, however, we will review some very simple, but important, parameters which characterize any finite population, but which are *not* to be used for the purpose of statistical inference. Ideas introduced in this section will be used again when we discuss sampling in Chapters 5 and 6.

3.1.1. Measures of Location

Definition 3.1. The *arithmetic mean* μ_a of a set of N values x_i $(i = 1, ..., N)$ is defined by

$$\mu_a = \frac{1}{N}\sum_{i=1}^{N} x_i. \tag{3.1}$$

Although the arithmetic mean is the most commonly used measure of location, and what is usually meant when one talks of the "mean", there exist other measures of location which in some circumstances are more useful. We will not dwell on these points here because in most physical work the arithmetic mean is (in a sense which will be considered later) the "best" measure of location. In fact, in later sections, we will revert to the usual convention of calling the arithmetic mean simply "the mean" and denote it by μ. However, below are given two other commonly used measures of location which are also used from time to time.

Definition 3.2. If the quantities $x_1, x_2, ..., x_N$ are arranged in increasing (or decreasing) value and then renumbered as $x_{(1)}, x_{(2)}, ..., x_{(N)}$, the *median* μ_m is defined as the middle value of the new set, for N odd, and as the mid-point of the middle pair of values if N is even.

DEFINITION 3.3. The *mode* M is that value of x in the set $x_1, x_2, \ldots, x_N$ which occurs with maximum frequency.

3.1.2. MEASURES OF DISPERSION

Just as in the case of measures of location, where there exist several different possible parameters which can be used to characterize a population, there is more than one possible measure of the *dispersion*, or scatter, of the measurements within the population. We shall simply mention the two most useful.

DEFINITION 3.4. The *mean deviation* δ_m is defined as the arithmetic mean of the absolute values of the deviations of the observations from the median

$$\delta_m = \frac{1}{N} \sum_{i=1}^{N} |x_i - \mu_m|. \tag{3.2}$$

DEFINITION 3.5. The *variance* σ^2 of a population is defined as the arithmetic mean of the squares of the deviations of x_i from the arithmetic mean μ_a

$$\sigma^2 = \frac{1}{N} \sum_{i=1}^{N} (x_i - \mu_a)^2. \tag{3.3}$$

The square root of the variance σ is called the *standard deviation.* The ratio σ/μ_a is sometimes called the *coefficient of variation.*

3.1.3. MOMENTS AND SKEWNESS

DEFINITION 3.6. The nth moment of a population about an arbitrary point $\bar{x}$ is defined as

$$\mu_n' = \frac{1}{N} \sum_{i=1}^{N} (x_i - \bar{x})^n. \tag{3.4}$$

If the point $\bar{x}$ is taken to be the mean μ (from now on taken to be the arithmetic mean), then the moments are called the *central moments* and are conventionally written without a prime.

From (3.4) we have

$$\mu_0' = 1; \quad \mu_1' = \mu - \bar{x} = d; \quad \mu_2' = \sigma^2 + d^2, \tag{3.5}$$

$$\mu_0 = 1; \quad \mu_1 = 0; \quad \mu_2 = \sigma^2. \tag{3.6}$$

The general relation between the two sets of moments is

$$\mu_k = \sum_{r=0}^{k} \binom{k}{r} \mu_{k-r}' (-\mu_1')^r, \tag{3.7}$$

with its inverse

$$\mu_k' = \sum_{r=0}^{k} \binom{k}{r} \mu_{k-r} (\mu_1')^r. \tag{3.8}$$

The importance of moments stems from the fact that a knowledge of the first few essentially determines the general characteristics of the distribution.

The *skewness*, or deviation from a symmetrical form, is defined in more than one way, but a common practice is to use the ratio

$$\beta_1 \equiv \mu_3^2/\mu_2^3, \tag{3.9}$$

since $\mu_3 = 0$ for a population distributed symmetrically about the mean. A measure of *kurtosis*, or degree of peaking, is, likewise, often taken to be the ratio

$$\beta_2 = \mu_4/\mu_2^2. \tag{3.10}$$

We shall see later that $\beta_2 = 3$ for the so-called "Normal Distribution" and this value is often taken as a standard.

3.2 Continuous univariate distributions

We have, on several occasions above, referred to "distributions" without defining the term. Such anticipation is difficult to avoid in a brief work of this kind, and the intuitive understanding of simple concepts before they are formally defined will be used more than once again. However, to proceed further we must consider more carefully what is meant by a distribution. This will be done in the present chapter, and will be followed, in Chapter 4, by an account of some of the simpler properties of the more important distributions encountered in practice.

In books on advanced statistics it is usual to avoid stating all definitions twice, once for continuous distributions and once for discrete, by using Stieltjes integration. However, since we shall be concerned with only a few distributions we shall merely repeat the steps which prove necessary.

Firstly, we shall introduce the concept of a random variable.

DEFINITION 3.7. A *random variable* is a function which can take on a definite value at every point in the sample space. Thus, if we have a sample space S with an associated probability function P, and a random variable X defined over the sample space, then to each point x_i of S we can assign a probability $P[x_i]$, and a definite numerical value $X(x_i)$ for the random variable.

The number of "heads" obtained by tossing two coins is an example of a random variable which can assume the discrete values 0, 1 or 2. Thus, if we distinguish between the two coins, the sample space consists of the points

$$x_1 = (H, H); \quad x_2 = (H, T); \quad x_3 = (T, H); \quad x_4 = (T, T)$$

and

$$X(x_1) = 2; \quad X(x_2) = 1; \quad X(x_3) = 1; \quad X(x_4) = 0.$$

If the coins are both "true", then we can calculate $P[x_i]$ as follows

$$P[X = 2] = P[x_1] = \tfrac{1}{4},$$
$$P[X = 1] = P[x_2 \cup x_3] = \tfrac{1}{2},$$
$$P[X = 0] = P[x_4] = \tfrac{1}{4}.$$

A random variable can also be continuous if it assumes a continuum of values.

DEFINITION 3.8. The continuous random variable x is said to have a *probability density function* (*or simply a density function*) $f(x)$ if it satisfies the following conditions:

(i) $f(x)$ is a single-valued non-negative real number for all real values of x,

(ii) $f(x)$ is normalized to unity

$$\int_{-\infty}^{\infty} f(x)\, dx = 1, \tag{3.11}$$

(iii) the probability with which x falls between any two real values a and b for which $a < b$ is given by

$$P[a \leqslant x \leqslant b] = \int_a^b dx\, f(x). \tag{3.12}$$

DEFINITION 3.9. *The cumulative distribution function* (*or simply the distribution function*) $F(x)$ of the continuous random variable x is defined by

$$F(x) = \int_{-\infty}^{x} dt\, f(t). \tag{3.13}$$

Since the probability that a member chosen at *random* (i.e. by a method which makes it equally likely that each member of the population will be chosen) from a distribution has a value x is just the density function $f(x)$, for this is the proportion of the population with this value, it follows from this definition that the probability of the member having a value $\leqslant x$ is the distribution function $F(x)$. It also follows from Eqn (3.13) that $F(x)$ is a non-decreasing function of x and that $0 \leqslant F(x) \leqslant 1$. This definition is clearly consistent with Definition 2.4 of Chapter 2. However, we should again note the element of circularity in the concept of randomness defined in terms of probability.

In terms of these formal definitions we may rewrite some of the earlier definitions of Chapter 2. Thus the mean about a point $\bar{x}$ is

$$\mu_{\bar{x}} = \int_{-\infty}^{\infty} dx\, f(x)(x - \bar{x}), \tag{3.14}$$

and the variance

$$\sigma^2 = \int_{-\infty}^{\infty} dx\, f(x)(x - \mu)^2, \tag{3.15}$$

both of which are special cases of the general moments

$$\mu_n' = \int_{-\infty}^{\infty} dx\, f(x)(x - \bar{x})^n, \tag{3.16}$$

and for $\bar{x} = \mu_1'$

$$\mu_n = \int_{-\infty}^{\infty} dx\, f(x)(x - \mu_1')^n. \tag{3.17}$$

The integrals in (3.16) and (3.17) may not converge for all n, and some distributions possess only the trivial zero-order moment ($\mu_0 = \mu_0' = 1$). In what follows we shall usually set $\bar{x} = 0$.

3.3 Expected values

The expected value of a random variable (or any function of a random variable) is obtained by finding the average value of the variable over all its possible values with due regard to the probability of their occurence. Our first definition follows from the definition of a density function.

DEFINITION 3.10. Let x be a continuous random variable with density function $f(x)$. Then the *expected value* of x, $E[x]$ is

$$E[x] = \int_{-\infty}^{\infty} x f(x)\, dx. \tag{3.18}$$

It can be proved from this definition that for a function $h(x)$ of x the expected value is

$$E[h(x)] = \int_{-\infty}^{\infty} dx\, h(x) f(x). \tag{3.19}$$

The following easily proved results hold for expected values involving a random variable x and a function $h(x)$, where c is a constant:

$$E[c] = c$$

$$E[ch(x)] = cE[h(x)] \tag{3.20}$$

$$E[h_1(x) + h_2(x)] = E[h_1(x)] + E[h_2(x)] \tag{3.21}$$

and

$$E[h_1(x_1)\, h_2(x_2)] = E[h_1(x_1)]\, E[h_2(x_2)] \tag{3.22}$$

if x_1, and x_2 are independent variates.

From Eqn (3.18) we see that the nth moment of a distribution about any point $\bar{x}$ is simply the expected value of $(x - \bar{x})^n$. Thus, for example,

$$\mu_n = E[(x - \mu_1')^n].$$

3.4 Generating functions and related topics

It was mentioned earlier that the usefulness of moments stems partly from the fact that a knowledge of them determines the form of the distribution function. This fact is embodied in the following theorem, which we shall state without proof.

THEOREM 3.1. *If the moments μ_n of a random variable x exist, and the series*

$$\sum_{n=1}^{\infty} \frac{\mu_n}{n!} r^n$$

converges absolutely for some $r > 0$, then the set of moments μ_n uniquely determines the distribution function.

In practice, a knowledge of the first few moments essentially determines the general characteristics of the distribution, and it is therefore worthwhile to construct a method which gives a representation of all the moments. Such a function is called the *moment generating function* (m.g.f.).

DEFINITION 3.11. If the random variable x has a density function $f(x)$ then the moment generating function (m.g.f.) is defined as

$$M_x(t) \equiv E[e^{xt}] = \int_{-\infty}^{\infty} dx\, e^{tx} f(x). \tag{3.23}$$

To generate the moments from (3.23) we expand exp (tx) giving

$$M_x(t) = E[1 + xt + \frac{1}{2!}(xt)^2 + \dots]$$

$$= \sum_{n=0}^{\infty} \frac{1}{n!} \mu_n' t^n. \tag{3.24}$$

Differentiating n times and setting $t = 0$ then gives

$$\mu_n' = \left. \frac{\partial^n M_x(t)}{\partial t^n} \right|_{t=0}. \tag{3.25}$$

In general, the m.g.f. about any point $\bar{x}$ is

$$M_{\bar{x}}(t) = E[\exp\{(x - \bar{x})t\}].$$

Thus, for $\bar{x} = \mu$ we have, using Eqn (3.20)

$$M_\mu(t) = e^{-\mu t} M_x(t). \tag{3.26}$$

Another important use of m.g.f.'s is in comparing two density functions, when the results of the following theorem may be used.

THEOREM 3.2. *Let x and y be two continuous random variables with density functions $f(x)$ and $g(y)$, respectively. If these distributions possess m.g.f.'s equal for some interval symmetric about the origin then $f(x) \equiv g(y)$.*

It is sometimes more convenient to consider, instead of the m.g.f., its logarithm. If we write a Taylor expansion for this quantity we have

$$\ln M_x(t) = \kappa_1 t + \kappa_2 \frac{t^2}{2!} + \dots,$$

where κ_n is the so-called *cumulant* of order n, and

$$\kappa_n = \left. \frac{\partial^n \ln M_x(t)}{\partial t^n} \right|_{t=0}.$$

The cumulants are rather simply related to the central moments of the distribution, the first few relations being

$$\kappa_1 = \mu_1$$

$$\kappa_2 = \mu_2$$

$$\kappa_3 = \mu_3$$

$$\kappa_4 = \mu_4 - 3\mu_2^2.$$

For some distributions the integral in Eqn (3.13) defining the m.g.f. will not exist, and in these circumstances another function called the *characteristic function* (c.f.) is introduced.

DEFINITION 3.12. If the random variable x has a density function $f(x)$ then the characteristic function (c.f.) is defined as

$$\phi_x(t) \equiv E[e^{itx}] = \int_{-\infty}^{\infty} dx\, e^{itx} f(x)$$
$$= M_x(it). \tag{3.27}$$

The characteristic function is very important in theoretical statistics, and a knowledge of it uniquely determines the density function. This result is known as the Inversion Theorem.

THEOREM 3.3 (The Inversion Theorem). *If $f(x)$ is a density function with a distribution function continuous everywhere and has a characteristic function $\phi_x(t)$ defined by Eqn* (3.27) *then*

$$f(x) = \frac{1}{2\pi}\int_{-\infty}^{\infty} dt\, \phi_x(t)\, e^{-ixt}. \tag{3.28}$$

Since $\int_{-\infty}^{\infty} f(x)\,dx$ is absolutely convergent, then, if $F(x)$ is continuous everywhere (as is required in Theorem 3.3), $\phi_x(t)$ is the Fourier transform of the density function $f(x)$, and the Inversion Theorem is simply the Fourier transform theorem.

3.5 Discrete univariate distributions

A probability density, and its associated distribution function, may be defined for a set of discrete values $x_1, x_2, \ldots, x_n$ by analogy with the definitions for continuous univariate distributions given in Section 3.2. Likewise, the results of Section 3.3 and 3.4 may be extended in a straightforward way to the case of discrete variables, and so we shall not discuss them further.

3.6 Multivariate distributions

The work of Sections 3.2–3.5 may also be extended to the case of multivariate distributions. We shall give below a few definitions for continuous variables. The results for discrete variables are obtainable by obvious substitutions.

DEFINITION 3.13. The n continuous random variables $x_1, x_2, \ldots, x_n$ are said to have a *multivariate joint density function* $f(x_1, x_2, \ldots, x_n)$ if,

(i) $f(x_1, x_2, \ldots, x_n)$ is a single-valued non-negative real number for all real values of $x_1, x_2, \ldots, x_n$,

(ii) $\int_{-\infty}^{\infty} \ldots \int_{-\infty}^{\infty} f(x_1, x_2, \ldots, x_n) \prod_{i=1}^{n} dx_i = 1,$

(iii) $P[a_1 \leqslant x_1 \leqslant b_1; \ldots; a_n \leqslant x_n \leqslant b_n]$

$$= \int_{a_n}^{b_n} \ldots \int_{a_1}^{b_1} f(x_1, \ldots, x_n) \prod_{i=1}^{n} dx_i,$$

where $P[a_1 \leqslant x_1 \leqslant b_1; \ldots; a_n \leqslant x_n \leqslant b_n]$ denotes the probability that x_1 falls between any two real number a_1 and b_1: *and* x_2 falls ...; *and* x_n falls ... a_n and b_n, *simultaneously*.

By analogy with the work of Chapter 2 we shall also define a marginal density function, and a conditional density function.

DEFINITION 3.14. If the n continuous random variables $x_i \; (i = 1, 2, \ldots, n)$ have a joint density function $f(x_1, x_2, \ldots, x_n) \equiv f(\mathbf{x})$, then the *marginal density function* of the variables $x_i \; (i = 1, 2, \ldots, m < n)$ is the value of $f(\mathbf{x})$ integrated with respect to all variables other than $x_1, x_2, \ldots, x_m$, i.e.

$$f^M(x_1, x_2, \ldots, x_m) \equiv \int_{-\infty}^{\infty} \ldots \int_{-\infty}^{\infty} f(x_1, \ldots, x_m, x_{m+1}, \ldots, x_n) \prod_{i=m+1}^{n} dx_i. \tag{3.29}$$

DEFINITION 3.15. The *multivariate conditional density* function of the random variables $x_i \; (i = 1, 2, \ldots, m < n)$, given the variables $x_{m+1}, x_{m+2}, \ldots, x_n$, is defined as

$$f^C(x_1, x_2, \ldots, x_m \mid x_{m+1}, x_{m+2}, \ldots, x_n) = \frac{f(x_1, x_2, \ldots, x_n)}{f^M(x_{m+1}, x_{m+2}, \ldots, x_n)}. \tag{3.30}$$

Again by analogy with the work of Chapter 2, this time Definition 2.10, we shall consider the concept of statistical independence.

DEFINITION 3.16. If the random variables $x_i \; (i = 1, 2, \ldots, n)$ may be split into groups such that the density function of the variables is expressible as a product of marginal density functions of the form

$$f(x_1, x_2, \ldots, x_n) = f_1{}^M(x_1, x_2, \ldots, x_i) f_2{}^M(x_{i+1}, x_{i+2}, \ldots, x_k) \ldots \ldots f_N{}^M(x_{l+1}, x_{l+2}, \ldots, x_n),$$

then the sets of variables

$$(x_1, \ldots, x_i);\ (x_{i+1}, \ldots, x_k);\ \ldots;\ (x_{l+1}, \ldots, x_n),$$

are said to be *statistically independent*, or *independently distributed.*

The *multivariate joint distribution function* $F(x_1, \ldots, x_n)$ of the n variables, $x_1, x_2, \ldots, x_n$ is obtained from the generalization of Definition 3.9. Thus

$$F(x_1, \ldots, x_n) = \int_{-\infty}^{x_n} \cdots \int_{-\infty}^{x_1} f(t_1, t_2, \ldots, t_n) \prod_{i=1}^{n} dt_i. \tag{3.31}$$

Likewise, the rth moment of the random variable x_i is obtained from the generalization of Definition 3.10. Thus

$$E[x_i^r] = \int_{-\infty}^{\infty} \cdots \int_{-\infty}^{\infty} x_i^r f(x_1, x_2, \ldots, x_n) \prod_{i=1}^{n} dx_i, \tag{3.32}$$

and from this relation we have

$$\mu_i = \int_{-\infty}^{\infty} \cdots \int_{-\infty}^{\infty} x_i f(x_1, x_2, \ldots, x_n) \prod_{i=1}^{n} dx_i, \tag{3.33}$$

and

$$\sigma_i^2 = \int_{-\infty}^{\infty} \cdots \int_{-\infty}^{\infty} (x_i - \mu_i)^2 f(x_1, x_2, \ldots, x_n) \prod_{i=1}^{n} dx_i. \tag{3.34}$$

Besides the individual moments defined by Eqn (3.32), the multiplicity of variables enables a number of *joint moments* to be defined. In general, these are given by

$$E[x_a^i x_b^j \ldots x_c^k] = \int_{-\infty}^{\infty} \cdots \int_{-\infty}^{\infty} (x_a^i x_b^j \ldots x_c^k) f(x_1, x_2, \ldots, x_n) \prod_{i=1}^{n} dx_i. \tag{3.35}$$

The most important joint moment is the covariance defined as follows.

DEFINITION 3.17. If the n random variables x_i $(i = 1, 2, \ldots, n)$ have a joint density function $f(x_1, x_2, \ldots, x_n)$ then the *covariance* of any two variables x_i and x_j is defined as

$$\operatorname{cov}(x_i, x_j) \equiv \sigma_{ij} = \int_{-\infty}^{\infty} \cdots \int_{-\infty}^{\infty} (x_i - \mu_i)(x_j - \mu_j) f(x_1, x_2, \ldots, x_n) \prod_{i=1}^{n} dx_i, \tag{3.36}$$

where μ_i and μ_j are given by Eqn (3.33). In terms of σ_{ij}, the *correlation coefficient* $\rho(x_i, x_j)$ is defined by

$$\rho(x_i, x_j) \equiv \frac{\text{cov}\,(x_i, x_j)}{\sigma(x_i)\,\sigma(x_j)}. \tag{3.37}$$

The correlation coefficient is a number lying between $+1$ and -1. It is a necessary condition for statistical independence that $\rho(x_i, x_j) = 0$. However, this is *not* a sufficient condition and $\rho(x_i, x_j) = 0$ does *not* imply that x_i and x_j are independently distributed. The following simple example illustrates this.

EXAMPLE 3.1. Let $x_1 = x$ and $x_2 = y = x^2$. Then

$$\begin{aligned}\text{cov}\,(x, y) &= E[xy] - E[x]\,E[y] \\ &= E[x^3] - E[x]\,E[x^2].\end{aligned}$$

Now if x has a density function which is symmetric about the mean then all the odd-order moments vanish and, in particular,

$$E[x] = E[x^3] = 0.$$

Thus $\text{cov}\,(x, y) = 0$ and hence $\rho(x, y) = 0$, even though x and y are not independent.

3.7 Functions of a random variable

In the previous sections of this chapter we have considered definitions relating to a continuous random variable x with a given density function $f(x)$. In practice, however, we may have occasion to refer to a function of x, e.g. $y(x)$, and the question arises: what is the density function of $y(x)$?

If $y = y(x)$ is monotonic (strictly increasing or decreasing) then the solution is simply

$$f(y\{x\}) = f(x\{y\}) \left| \frac{dx}{dy} \right|. \tag{3.38}$$

However, if $y(x)$ has a continuous non-zero derivative at all but a finite number of points we must split the range into a finite number of sections in each of which $y(x)$ *is* a monotonic strictly increasing or strictly decreasing function of x with a continuous derivative, and apply Eqn (3.38) to each section separately. Thus, at all points where (i) $dy/dx \neq 0$ *and* (ii) $y = y(x)$

has a real finite solution for $x = x(y)$, the required density function is

$$f(y\{x\}) = \prod_{\text{all } x} f(x\{y\}) \left| \frac{dy}{dx} \right|^{-1}. \tag{3.39}$$

If the above conditions are violated then $f(y\{x\}) = 0$ at that point.

EXAMPLE 3.2. Given a random variable x with density function

$$f(x) = \frac{1}{\sqrt{(2\pi)}} \exp\left(\frac{-x^2}{2}\right),$$

what is the density function of $y = x^2$?

Now

$$x = \pm\sqrt{y} \quad \text{and} \quad \frac{dy}{dx} = 2x = \pm 2\sqrt{y}.$$

Thus, for $y < 0$, x is not real and so

$$f(y\{x\}) = 0, \qquad y < 0.$$

For $y = 0$, $dx/dy = 0$ and so, again

$$f(y\{x\}) = 0, \qquad y = 0.$$

Finally, for $y > 0$ we may split the range into two parts, $x > 0$, and $x < 0$. Then, applying Eqn (3.39) gives

$$f(y\{x\}) = \frac{1}{2\sqrt{y}} [f(x = -\sqrt{y}) + f(x = +\sqrt{y})],$$

i.e.

$$f(y\{x\}) = \frac{1}{\sqrt{(2\pi y)}} \exp\left(\frac{-y}{2}\right).$$

Similar arguments to those above hold for the multivariate case. Thus, if we have n random variables y_i $(i = 1, 2, \ldots, n)$ which are themselves functions of the n random variables x_i $(i = 1, 2, \ldots, n)$ defined by

$$y_i = y_i(x_1, x_2, \ldots, x_n), \qquad i = 1, 2, \ldots, n$$

such that $\partial y_i/\partial x_j$ are continuous for all x_j, and such that the Jacobian

$$J = \det\left\{\frac{\partial y_i}{\partial x_j}\right\} \neq 0,$$

then, in the region where there is a unique solution for x_j in terms of the y_i we have

$$f(y_1, y_2, \ldots, y_n) = |J|^{-1} f(x_1, x_2, \ldots, x_n). \tag{3.40}$$

Again, if the above conditions do not hold then the range of the variables can always be split into sections, as for the univariate case, and Eqn (3.39) applied in each section.

EXAMPLE 3.3. Given the two random variables x_1 and x_2 with a joint density function

$$f(x_1, x_2) = \frac{1}{2\pi} \exp\{-\tfrac{1}{2}(x_1^2 + x_2^2)\},$$

what is the density of the variables

$$y_1 = x_1/x_2 \quad \text{and} \quad y_2 = x_1?$$

The Jacobian of the transformation is

$$J = \det\begin{pmatrix} 1/x_2 & -x_1/x_2^2 \\ 1 & 0 \end{pmatrix}$$

$$= \frac{x_1}{x_2^2} = \frac{y_1^2}{y_2}.$$

Thus, applying (3.39), (*provided* $J \neq 0$),

$$f(y_1, y_2) = \frac{1}{2\pi} \frac{|y_2|}{y_1^2} \exp\left\{-\tfrac{1}{2}\left(y_2^2 + \frac{y_2^2}{y_1^2}\right)\right\}, \qquad (y_1 \neq 0).$$

4 Theoretical Distributions: Examples

In Chapter 3 we have considered the general properties of theoretical distributions. In this chapter we will consider the forms and properties of some specific distributions commonly encountered in practice, and two others which are useful for illustrative purposes.

4.1 Uniform distribution

DEFINITION 4.1. The *uniform distribution* for a continuous variable x has a density function

$$f(x) \equiv u(x; c, d) = \begin{cases} \dfrac{1}{d-c} & c \leqslant x \leqslant d \\ 0 & \text{otherwise} \end{cases} \tag{4.1}$$

The distribution function obtained from (4.1) is

$$F(x) = \begin{cases} 0 & x < c \\ \dfrac{x-c}{d-c} & c \leqslant x \leqslant d \\ 1 & x > d \end{cases} \tag{4.2}$$

From (3.14) and (3.15) we can easily show that the mean and variance are given by

$$\mu = \frac{c+d}{2}; \qquad \sigma^2 = \frac{d-c}{12}. \tag{4.3}$$

The Dirac δ-function, well-known to physicists, may be considered as the limiting form of a uniform distribution where $d \to c$.

Although the uniform distribution is the simplest of all continuous distributions it is useful in practice for studying rounding errors in measurements made within a given accuracy. However, its theoretical significance is enhanced by the following easily proved theorem.

THEOREM 4.1. *Let $f(x)$ be any density function of a continuous random variable x, and let $F(x)$ be its distribution function. Then $f(x)$ may be transformed to the uniform density function*

$$g(u) = 1, \qquad 0 \leqslant u \leqslant 1$$

by the transformation $u = F(x)$.

By using this theorem it is possible to exhibit many properties of continuous distributions in general, by proving them for the paricular case of the uniform distribution. It also follows from this theorem that there exists at least one transformation which transforms any continuous distribution into any other, it being simply the product of the transformations which take each distribution into the uniform distribution.

4.2 Univariate normal distribution

This distribution is by far the most important in statistics since many distributions encountered in practice are believed to be of approximately normal form, a point to which we will return in Chapter 5.

DEFINITION 4.2. The *normal density function* for a continuous random variable x is defined to be

$$f(x) \equiv n(x;\, \mu, \sigma) = \frac{1}{\sigma}\frac{1}{(2\pi)^{\frac{1}{2}}}\exp\left[-\tfrac{1}{2}\left(\frac{x-\mu}{\sigma}\right)^2\right], \tag{4.4}$$

and its distribution function is

$$F(x) \equiv N(x;\, \mu, \sigma) = \frac{1}{\sigma}\frac{1}{(2\pi)^{\frac{1}{2}}}\int_{-\infty}^{x} dt \exp\left[-\tfrac{1}{2}\left(\frac{t-\mu}{\sigma}\right)^2\right]. \tag{4.5}$$

$F(x)$ is also called a *Gaussian* distribution function. Graphs of $f(x)$ and $F(x)$ for $\mu = 0$ and $\sigma = 0{\cdot}5$, $1{\cdot}0$ and $2{\cdot}0$ are shown in Fig. (4.1).

Since this is the first non-trivial distribution we have encountered it will be useful to implement some of our previous definitions. Firstly, it is clear from Eqn (4.4) that $f(x)$ is a single-valued non-negative real number for all values of $x (\sigma > 0)$. Furthermore, by the transformation

$$t^2 = \tfrac{1}{2}\left(\frac{x-\mu}{\sigma}\right)^2,$$

we can write

$$\int_{-\infty}^{\infty} dx\, f(x) = \frac{1}{\pi^{\frac{1}{2}}} \int_{-\infty}^{\infty} dt\, e^{-t^2}.$$

Since the latter integral is $(\pi)^{\frac{1}{2}}$ we see that $f(x)$ is normalized to unity and is thus a valid density function.

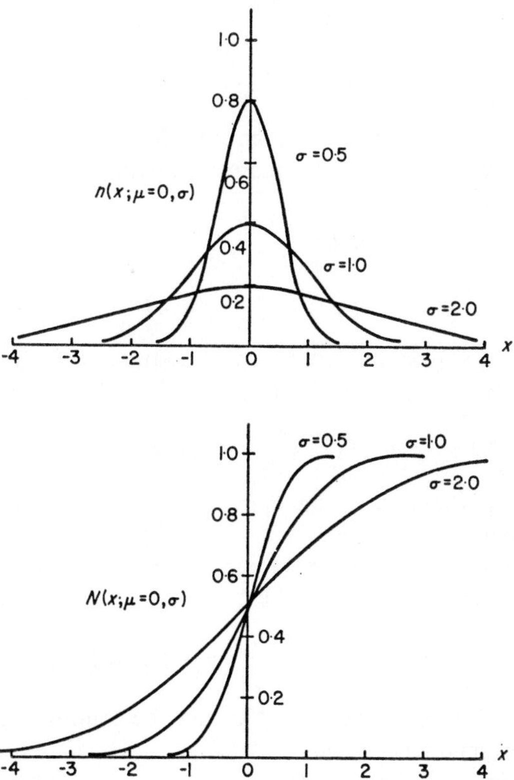

FIG. 4.1. The normal density $n(x;\, \mu, \sigma)$, and its distribution function $N(x;\, \mu, \sigma)$, for $\mu = 0$ and $\sigma = 0{\cdot}5$, $1{\cdot}0$ and $2{\cdot}0$.

To find the moments of the normal distribution we first find the m.g.f. From Definition 3.11, Eqn (3.23),

$$M_x(t) = E[\exp(tx)] = \exp(t\mu)\, E[\exp\{t(x-\mu)\}]$$

$$= \frac{\exp(t\mu)\exp(\sigma^2 t^2/2)}{(2\pi)^{\frac{1}{2}}\sigma} \int_{-\infty}^{\infty} dx \exp\left[\frac{(x-\mu-\sigma^2 t)^2}{-2\sigma^2}\right].$$

This integral is related to the area under a normal curve with mean $(\mu + \sigma^2 t)$ and variance σ^2. Thus

$$M_x(t) = \exp(t\mu + \sigma^2 t^2/2). \tag{4.6}$$

On differentiating (4.6) twice and setting $t = 0$ we have

$$\mu_1' = \mu$$

$$\mu_2' = \sigma^2 + \mu^2$$

and

$$\operatorname{var}(x) = \mu_2' - (\mu_1')^2 = \sigma^2.$$

Thus the mean end variance of the normal distribution are μ and σ^2, respectively. The same techniques for moments about the mean give

$$\begin{aligned} \mu_{2n} &= \frac{(2n)!}{n!2^n}\sigma^{2n}, \\ \mu_{2n+1} &= 0, \qquad n \geqslant 1. \end{aligned} \tag{4.7}$$

The odd-order moments are zero by virtue of the symmetry of the distribution. Using (4.7) in Eqns (3.9) and (3.10) gives

$$\beta_1 = 0; \qquad \beta_2 = 3. \tag{4.8}$$

This value of β_2 is taken as a standard against which the kurtosis of other distributions may be compared.

Using essentially the same technique as was used to derive the m.g.f. above we can show that the normal c.f. is

$$\phi(t) = \exp[it\mu - t^2\sigma^2/2], \tag{4.9}$$

which agrees with Eqn (3.27), and which may be confirmed by applying the result of the Inversion Theorem.

For the special case when $\mu = 0$ and $\sigma^2 = 1$ we have, from (4.4) and (4.5)

$$n(t; 0; 1) = \frac{1}{(2\pi)^{\frac{1}{2}}}\exp(-t^2/2), \tag{4.10}$$

$$N(t; 0; 1) = \frac{1}{(2\pi)^{\frac{1}{2}}}\int_{-\infty}^{t} du \exp(-u^2/2). \tag{4.11}$$

These forms are called the *standard normal density function*, and *standard normal distribution function*, respectively, and will simply be denoted by

$n(t)$ and $N(t)$. Tables of $n(t)$ and $N(t)$ are given in Appendix D. They may be used for all families of normal curves if the following, easily proved, relations are noted:

$$n(-t) = n(t), \tag{4.12}$$

$$N(-t) = 1 - N(t), \tag{4.13}$$

$$2\int_0^t n(u)\,du = \int_{-t}^t du\,n(u) = 2N(t) - 1. \tag{4.14}$$

Using Eqns (4.12) – (4.14) and the tables in Appendix D the following useful results may be deduced.

(i) The proportion of standard normal values contained within 1, 2 and 3 standard deviations from the mean are 68·3%, 95·4%, and 99·7%, respectively.

(ii) If t_α denotes that value of the standard normal distribution for which

$$\int_{t_\alpha}^{\infty} dt\,n(t; 0; 1) = \alpha, \tag{4.15}$$

then $(\mu \pm t_\alpha\sigma)$ are the limits of the $100(1-2\alpha)\%$ symmetric interval about μ.

The usefulness of such results will be evident when we discuss confidence intervals in Chapter 10.

One final result which we shall quote for the univariate normal concerns the distribution of a linear sum of normally distributed random variables.

THEOREM 4.2. *If x_i $(i = 1, 2, \ldots, n)$ are n independent random variables having normal distributions $N(x_i; \mu_i, \sigma_i^2)$ then the random variable $T = \sum_i a_i x_i$ is distributed as $N(T; \mu, \sigma^2)$ where*

$$\mu = \sum_{i=1}^{n} a_i\mu_i \quad \textit{and} \quad \sigma^2 = \sum_{i=1}^{n} a_i^2\sigma_i^2.$$

Proof. The c.f. of T is given by

$$\begin{aligned}\phi_T(t) &= E[\exp(itT)] \\ &= E\left[\exp\left(it\sum_{i=1}^{n} a_i x_i\right)\right].\end{aligned}$$

Using the fact that the x_i are independent we may write this as [*cf.* Eqn (3.22)]

$$\phi_T(t) = \prod_{i=1}^{n} E[\exp(ita_i x_i)]$$

$$= \prod_{i=1}^{n} \phi_i(t),$$

where $\phi_i(t)$ is the c.f. of the random variable $(a_i x_i)$. Now we have previously shown that [*cf.* Eqn (4.9)]

$$\phi_i(t) = \exp[ita_i\mu_i - t^2\sigma_i^2 a_i^2/2],$$

and so

$$\phi_T(t) = \exp\left\{\sum_{i=1}^{n} (ita_i\mu_i - t^2 a_i^2\sigma_i^2/2)\right\},$$

i.e.

$$\phi_T(t) = \exp(it\mu - t^2\sigma^2/2),$$

where

$$\mu = \sum_{i=1}^{n} a_i\mu_i; \qquad \sigma^2 = \sum_{i=1}^{n} a_i^2\mu_i^2.$$

But this is the c.f. of a normal variate whose mean is μ and whose variance is σ^2. Thus, by the Inversion Theorem, T is distributed as $N(T; \mu, \sigma^2)$.

4.3 Multivariate normal distribution

The normal distribution for the multivariate case is defined as follows:

DEFINITION 4.3. If $x_1, x_2, \ldots, x_n \equiv \mathbf{x}$ are n random variables, then the *multivariate normal density function*, of order n, is

$$f(\mathbf{x}; \boldsymbol{\mu}, \mathbf{V}) = \frac{1}{(2\pi)^{n/2}|\mathbf{V}|^{\frac{1}{2}}} \exp[-\tfrac{1}{2}(\mathbf{x} - \boldsymbol{\mu})^T \mathbf{V}^{-1}(\mathbf{x} - \boldsymbol{\mu})], \qquad (4.16)$$

where $\boldsymbol{\mu}$ is a constant vector, which is the mean of the distribution, and $\mathbf{V}$ is a symmetric positive–definite matrix, which is the variance matrix of the vector $\mathbf{x}$. The quantity

$$Q = (\mathbf{x} - \boldsymbol{\mu})^T \mathbf{V}^{-1}(\mathbf{x} - \boldsymbol{\mu}), \qquad (4.17)$$

is called the *quadratic form* of the multivariate normal distribution.

The multivariate normal distribution possesses a number of important properties, and we shall consider three of these here. The first concerns the form of the joint marginal distribution of a subset of the n variables.

THEOREM 4.3. *If the n random variables* $x_1, x_2, \ldots, x_n$ *are distributed as the n-variate normal distribution then the joint marginal distribution of any subset* x_i $(i = 1, 2, \ldots, m < n)$ *is the m-variate normal.*

This theorem can be proved in a straightforward manner by simply constructing the joint marginal distribution from Eqn (4.16), but we will not reproduce the details here. It follows from Theorem 4.3 that the distribution of any single random variable in the set x_i (this is the case $m = 1$) is distributed as the univariate normal. We shall use this result in the second theorem which concerns the conditions under which the variables of the distribution are independent.

THEOREM 4.4. *If the random variables* $x_1, x_2, \ldots, x_n \equiv \mathbf{x}$ *have the multivariate normal distribution with mean vector* $\boldsymbol{\mu}$ *and variance matrix* $\mathbf{V}$, *then the components of* $\mathbf{x}$ *are jointly independent if, and only if,* $\text{cov}(x_i, x_j) = 0$ *for all* $i \neq j$.

Proof. If $\text{cov}(x_i, x_j) = 0$ for $i \neq j$ then the variance matrix $\mathbf{V}$ is diagonal. In this case the quadratic form of the distribution becomes

$$(\mathbf{x} - \boldsymbol{\mu})^T\mathbf{V}^{-1}(\mathbf{x} - \boldsymbol{\mu}) = \sum_{i=1}^{n} (x_i - \mu_i)^2 V_{ii}^{-1},$$

and so the density function may be written

$$f(\mathbf{x}) = \frac{1}{(2\pi)^{n/2}|\mathbf{V}|^{\frac{1}{2}}} \exp[-\tfrac{1}{2}(\mathbf{x} - \boldsymbol{\mu})^T\mathbf{V}^{-1}(\mathbf{x} - \boldsymbol{\mu})]$$

$$= \frac{1}{(2\pi)^{n/2}} \prod_{i=1}^{n} (V_{ii})^{-\frac{1}{2}} \exp\left[-\tfrac{1}{2}\sum_{i=1}^{n} (x_i - \mu_i)^2 V_{ii}^{-1}\right],$$

i.e.

$$f(\mathbf{x}) = \prod_{i=1}^{n} f_i(x_i),$$

where

$$f_i(x_i) = \frac{1}{(2\pi)^{\frac{1}{2}}} \frac{1}{V_{ii}^{\frac{1}{2}}} \exp\left[-\tfrac{1}{2}\frac{(x_i - \mu_i)^2}{V_{ii}}\right]. \tag{4.18}$$

Now (4.18) is the form of the density function for a univariate normal distribution and so, by virtue of Theorem 4.3 and Definition 3.16, the variables x_i are independently distributed.

To establish the inverse, i.e. that if the x_i are jointly independent then $\mathbf{V}$ is diagonal, we start from the definition of $\mathbf{V}$.

$$V_{ij} = \text{cov}(x_i, x_j) = E[(x_i - \mu_i)(x_j - \mu_j)], \ (i \neq j)$$

$$= \int_{-\infty}^{\infty} \cdots \int_{-\infty}^{\infty} (x_i - \mu_i)(x_j - \mu_j) f(\mathbf{x}) \prod_{k=1}^{n} dx_k.$$

Now since x_i and x_j are independent we have

$$f(\mathbf{x}) = \prod_{i=1}^{n} f_i(x_i),$$

and hence

$$V_{ij} = \int_{-\infty}^{\infty} (x_i - \mu_i) f_i(x_i)\, dx_i \int_{-\infty}^{\infty} (x_j - \mu_j) f_j(x_j)\, dx_j$$

$$\prod_{k \neq i,j}^{n} \int_{-\infty}^{\infty} f_k(x_k)\, dx_k.$$

But, by definition,

$$\int_{-\infty}^{\infty} (x_i - \mu_i) f_i(x_i)\, dx_i = 0,$$

and so,

$$V_{ij} = 0 \quad \text{for all} \quad i \neq j.$$

This completes the proof.

The third, and final, result concerns the distribution of linear combinations of random variables, each of which itself has a univariate normal distribution. The following theorem is a generalization of Theorem 4.2.

THEOREM 4.5. *If* $\mathbf{x} \equiv x_1, x_2, \ldots, x_n$ *has a multivariate normal distribution with mean* $\boldsymbol{\mu}$ *and variance matrix* $\mathbf{V}$, *then any linear combination of* x_i, *say*

$$S = \sum_{i=1}^{n} a_i x_i,$$

with a_i *a set of constants, has a univariate normal distribution with mean*

$$\mu = \sum_{i=1}^{n} a_i \mu_i,$$

and variance

$$\sigma^2 = \sum_{j=1}^{n} \sum_{i=1}^{n} a_i a_j V_{ij}.$$

Proof. Let

$$S = \sum_{i=1}^{n} a_i x_i = \mathbf{x}^T \mathbf{A},$$

where $\mathbf{A} = \{a_i\}$. The m.g.f. of S is

$$\begin{aligned} M_S(t) &= E[\exp(St)] \\ &= E[\exp(\mathbf{x}^T\mathbf{A})t] \\ &= E[\exp\{(\mathbf{x} - \boldsymbol{\mu})^T\mathbf{A}t + (\boldsymbol{\mu}^T\mathbf{A})t\}] \\ &= \exp[(\boldsymbol{\mu}^T\mathbf{A})t]\, E[\exp(\mathbf{x} - \boldsymbol{\mu})^T\mathbf{A}t]. \end{aligned}$$

Now

$$E[\exp(\mathbf{x} - \boldsymbol{\mu})^T\mathbf{A}t] = \exp\left[\frac{t^2}{2}(\mathbf{A}^T\mathbf{V}\mathbf{A})\right],$$

thus

$$M_S(t) = \exp[(\boldsymbol{\mu}^T\mathbf{A})\,t + (\mathbf{A}^T\mathbf{V}\mathbf{A})\,t^2/2]. \tag{4.19}$$

But from Eqn (4.6) this is the m.g.f. of a normal variate with mean

$$\mu = \boldsymbol{\mu}^T\mathbf{A} = \sum_{i=1}^{n} a_i \mu_i,$$

and

$$\text{variance } \sigma^2 = \mathbf{A}^T\mathbf{V}\mathbf{A} = \sum_{i=1}^{n} \sum_{j=1}^{n} a_i a_j V_{ij}.$$

Thus, by Theorem 3.2, S is distributed as $N(S, \mu, \sigma^2)$.

An important special example of the multivariate normal distribution is the bivariate case, which occurs frequently in practice. The density function is

$$n(x, y;\, \mu_x, \mu_y, \sigma_x, \sigma_y, \rho) \equiv n(x, y) = \frac{1}{2\pi\sigma_x\sigma_y(1-\rho^2)^{\frac{1}{2}}} \exp\left[\frac{-R}{2(1-\rho^2)}\right], \tag{4.20}$$

where

$$R = \left(\frac{x - \mu_x}{\sigma_x}\right)^2 - 2\left(\frac{x - \mu_x}{\sigma_x}\right)\left(\frac{y - \mu_y}{\sigma_y}\right) + \left(\frac{y - \mu_y}{\sigma_y}\right)^2, \tag{4.21}$$

and ρ is the correlation coefficient, as defined by Eqn (3.37). If the exponent in (4.20) is a constant $(-K)$, i.e.

$$R = 2(1 - \rho^2)K,$$

then the points (x, y) lie on an ellipse with centre (μ_x, μ_y). In fact the density function (4.20) is a bell-shaped surface, and any plane parallel to the xy plane which cuts this surface will intersect it in an elliptical curve. Any plane perpendicular to the xy plane will cut the surface in a curve of the normal form.

Just as for the univariate normal distribution we can define a *standard bivariate normal density function*

$$n(u, v) = \frac{1}{2\pi(1 - \rho^2)^{\frac{1}{2}}} \exp\left[\frac{(u^2 - 2\rho uv + v^2)}{-2(1 - \rho^2)}\right]. \tag{4.22}$$

A feature of the bivariate normal distribution is that for $\rho = 0$

$$n(u, v) = n(u)n(v), \tag{4.23}$$

which from Theorem 4.4 implies that u and v are independently distributed, a result which is *not* generally true for all bivariate density functions.

Finally, the joint moment generating function for this distribution may be obtained as follows

$$\begin{aligned} M_{xy}(t_1, t_2) &= E[\exp(t_1 x + t_2 y)] \\ &= \int_{-\infty}^{\infty}\int_{-\infty}^{\infty} \exp(t_1 x + t_2 y) f(x, y)\, dx\, dy. \end{aligned} \tag{4.24}$$

Using the same technique which we used for obtaining the m.g.f. for the univariate normal distribution we set

$$u = \frac{x - \mu_x}{\sigma_x}; \qquad v = \frac{y - \mu_y}{\sigma_y}.$$

Then

$$M_{xy}(t_1, t_2) = \frac{\exp(t_1\mu_x + t_2\mu_y)}{2\pi(1 - \rho^2)^{\frac{1}{2}}} \iint e^{(t_1\sigma_x u + t_2\sigma_y v)} \exp\left[\frac{u^2 - 2\rho uv + v^2}{-2(1 - \rho^2)}\right] du\, dv,$$

and substituting

$$w = \frac{u - \rho v - (1 - \rho^2)t_1\sigma_x}{(1 - \rho^2)^{\frac{1}{2}}},$$

$$z = v - \rho t_1\sigma_x - t_2\sigma_y,$$

we have

$$M_{xy}(t_1, t_2) = \exp\left[t_1\mu_x + t_2\mu_y + \tfrac{1}{2}(t_1^2\sigma_x^2 + 2\rho t_1 t_2\sigma_x\sigma_y + t_2^2\sigma_y^2)\right]. \qquad (4.25)$$

The moments may be obtained in the usual way by evaluating the derivatives of Eqn (4.25) at $t_1 = t_2 = 0$. Thus, e.g.

$$E[x^2] = \left.\frac{\partial^2 M_{xy}(t_1, t_2)}{\partial t_1^2}\right|_{t_1=t_2=0}$$

$$= \sigma_x^2 + \mu_x^2.$$

4.4 Cauchy distribution

The Cauchy distribution is a simple form occurring frequently in practice (i.e. the shape of atomic spectral lines) but possessing properties which serve as a useful reminder that not all distributions are well-behaved. It is defined as follows.

DEFINITION 4.4. The *Cauchy density function* for a random variable x is

$$f(x;\theta) = \frac{1}{\pi}\cdot\frac{1}{1 + (x - \theta)^2}, \qquad -\infty < x < \infty.$$

The parameter θ can be interpreted as the mean μ of the distribution only if the definition of the mean is extended as follows,

$$\mu = \lim_{N\to\infty}\int_{-N}^{N} dx\, f(x;\theta)x.$$

This is somewhat questionable and we will, in general, set $\theta = 0$. Then the distribution function becomes

$$F(x) = \frac{1}{2} + \frac{1}{\pi}\arctan(x).$$

The moment about the mean (zero) of order $2n$ is

$$\mu_{2n} = \frac{1}{\pi}\int_{-\infty}^{\infty} dx \frac{x^{2n}}{1 + x^2}, \qquad (4.26)$$

but this integral converges only for $n = 0$, and so only the trivial moment $\mu_0 = 1$ exists. Finally, it can be shown that the ratio of two standardized normal deviates has a Cauchy density function. This is one reason why it is encountered in practice.

4.5 Binomial distribution

Consider a population of members each of which either possesses a certain attribute P, or does not possess this attribute. Denote the latter possibility by Q. If the proportion of members possessing P is p, and that possessing Q is q, then clearly $(p + q) = 1$. An experiment involving such a population is called a *Bernoulli trial*, i.e. one with only two possible outcomes. Suppose now we wish to choose sets from the population, each of which contains n members. The proportion of cases containing rP's and $(n - r)Q$'s will be

$$\binom{n}{r} p^r q^{n-r}, \tag{4.27}$$

i.e. the rth term in the binomial expansion of

$$f(p, q) = (q + p)^n. \tag{4.28}$$

Expressed in another way, if p is the chance of an event happening in a single trial, then for n independent trials the terms in the expansion

$$f(p, q) = q^n + nq^{n-1}p + \ldots + p^n,$$

give the chances of 0, 1, 2, ..., n events happening. Thus we are led to the following definition.

DEFINITION 4.5. The probability density of the *binomial distribution* is defined as

$$f(r; p, q) = \binom{n}{r} p^r q^{n-r}, \tag{4.29}$$

and gives the probability of obtaining $r = 0, 1, 2, \ldots, n$ successes, i.e. events having the attribute P, in an experiment consisting of n Bernoulli trials (i.e. $p + q = 1$).

Tables of the binomial distribution function are given in Appendix D.

The moment generating function may be found directly from Eqn (4.29) and the definition (3.23), which for the discrete case becomes

$$M_r(t) = \sum_{r=0}^{n} f(r; p, q)\, e^{tr}.$$

Using (4.29) we have

$$M_r(t) = \sum_{r=0}^{n} \binom{n}{r} p^r q^{n-r} \mathrm{e}^{tr}$$

$$= (p\,\mathrm{e}^t + q)^n, \tag{4.30}$$

and hence, using Eqn (3.25), we have

$$\mu_1' = \mu = np,$$

$$\mu_2' = np + n(n-1)p^2, \tag{4.31}$$

and

$$\sigma^2 = \mu_2' - (\mu_1')^2 = npq. \tag{4.32}$$

The m.g.f. for moments about the mean is

$$M_\mu(t) = \mathrm{e}^{-\mu t} M(t), \tag{4.33}$$

and gives

$$\mu_3 = npq(q-p),$$

$$\mu_4 = npq[1 + 3(n-2)pq]. \tag{4.34}$$

Using (4.34) in Eqns (3.9) and (3.10) gives

$$\beta_1 = (q-p)^2/(npq),$$

$$\beta_2 = 3 + (1 - 6pq)/(npq), \tag{4.35}$$

which tend to the values for the normal distribution as $n \to \infty$. It is, in fact, of interest to consider the limiting form of the binomial distribution as $n \to \infty$, and this is provided by the following theorem.

THEOREM 4.6. *The limiting form of the binomial distribution as $n \to \infty$ is the standard form of the normal distribution.*

Proof. The characteristic function of the binomial distribution is, from Eqns (3.27) and (4.30),

$$\phi_r(t) = (q + p\,\mathrm{e}^{it})^n. \tag{4.36}$$

Now any distribution may be expressed in standard measure (i.e. with $\mu = 0$ and $\sigma^2 = 1$) by the transformation

$$x = (r - \mu)/\sigma, \tag{4.37}$$

and so, from the definition of a c.f., we have

$$\phi_r(t) = \int_{-\infty}^{\infty} dx\, f(x) \exp[it(\sigma x + \mu)]$$

$$= \exp(it\mu)\, \phi_x(\sigma t).$$

From Eqns (4.31) and (4.32)

$$\mu = np \quad \text{and} \quad \sigma^2 = npq.$$

Thus, using Eqn (4.36) we have

$$\phi_x(t) = \exp\left[\frac{-itnp}{(npq)^{\frac{1}{2}}}\right]\left\{q + p\exp\left[\frac{it}{(npq)^{\frac{1}{2}}}\right]\right\}^n,$$

giving

$$\ln \phi_x(t) = \frac{-itnp}{(npq)^{\frac{1}{2}}} + n\ln\left\{1 + p\left[\exp\left(\frac{it}{(npq)^{\frac{1}{2}}}\right) - 1\right]\right\},$$

where we have used the relation $p + q = 1$ in the log term. If we now let $n \to \infty$, keeping t finite, we may expand the log term and we find

$$\ln \phi_x(t) = -\,t^2/2 + O(t^3 n^{-\frac{1}{2}}).$$

Thus, for any finite t

$$\phi(t) \to \exp(-\,t^2/2).$$

However, this is the form of the c.f. of a standardized normal distribution [*cf.* Eqn (4.9)], and, by the Inversion Theorem, the associated density function is

$$f(x) = \frac{1}{(2\pi)^{\frac{1}{2}}} \exp(-\,x^2/2), \tag{4.38}$$

which is the standardized form of the normal distribution.

The normal approximation to the binomial distribution is reasonable even down to values of $n \sim 8$.

4.6 Multinomial distribution

The multinomial distribution is the generalization of the binomial distribution to the case of repeated trials where there are more than two possible outcomes. It is defined as follows

DEFINITION 4.6. If an event may occur with k possible outcomes each with a probability p_i ($i = 1, 2, \ldots, k$),

$$\sum_{i=1}^{k} p_i = 1, \tag{4.39}$$

and if r_i is the number of times the outcome associated with p_i occurs, then the density function for the random variables r_i ($i = 1, 2, \ldots, k-1$) is the *multinomial* and is defined as

$$f(r_1, r_2, \ldots, r_{k-1}) = \frac{n!}{\prod_{i=1}^{k} r_i!} \prod_{i=1}^{k} p_i^{r_i}, \qquad r_i = 0, 1, \ldots, n\,. \tag{4.40}$$

Note that each of the r_i may range from 0 to n inclusive, and that only $(k-1)$ variables are involved because of the linear constraint

$$\sum_{i=1}^{k} r_i = n.$$

With suitable generalizations the results of Section 4.5 may be extended to the multinomial distribution, and, in particular, this distribution tends, in the limit, to the multivariate normal distribution.

4.7 Poisson distribution

The Poisson distribution is an important distribution occurring frequently in practice which is derived from the binomial distribution by a special limiting process. Consider the binomial distribution for the case when p, the proportion of the population possessing the attribute P, is very small but n, the number of members of a given set, is large such that

$$\lim_{p\to 0} (np) = r, \tag{4.41}$$

where r is a finite positive constant, i.e. where

$$n \gg np \gg p.$$

The kth term in the binomial distribution then becomes

$$\lim_{p\to 0} \left[\binom{n}{k} p^k q^{n-k}\right] = \frac{r^k}{k!} \exp(-r).$$

This is the density function of the Poisson distribution.

DEFINITION 4.7. The density function of the Poisson distribution is

$$f(k;r) = \frac{e^{-r}r^k}{k!}, \qquad r > 0, \quad k = 0, 1, \ldots \tag{4.42}$$

and gives the probability for different events when the chance of an event is small but the total number of trials is large. Although, in principle, the number of values of k is infinite the rapid convergence of successive terms in (4.42) means that, in practice, the distribution function is accurately given by the first few terms. Tables of the Poisson distribution function are given in Appendix D.

The m.g.f. for the Poisson distribution is

$$\begin{aligned} M_k(t) &= E[e^{kt}] \\ &= \sum_{k=0}^{\infty} \frac{e^{kt}e^{-r}r^k}{k!} = e^{-r} \sum_{k=0}^{\infty} \frac{(r\,e^t)^k}{k!} \\ &= e^{-r}\exp[r\,e^t]. \end{aligned} \tag{4.43}$$

Differentiating (4.43) and setting $t = 0$ gives

$$\begin{aligned} \mu_1' &= r \\ \mu_2' &= r(r+1) \\ \mu_3' &= r[(r+1)^2 + r] \\ \mu_4' &= r[r^3 + 6r^2 + 7r + 1], \end{aligned} \tag{4.44}$$

and from (3.7)

$$\begin{aligned} \mu_2 &= r \\ \mu_3 &= r \\ \mu_4 &= r(3r+1). \end{aligned} \tag{4.45}$$

Thus,

$$\mu = \sigma^2 = r, \tag{4.46}$$

a simple result which is very useful in practice. Also from (4.45), (3.9) and (3.10), we have

$$\beta_1 = \frac{1}{r}; \qquad \beta_2 = 3 + \frac{1}{r}. \tag{4.47}$$

From these results on the skewness parameters one might suspect that as $r \to \infty$ the Poisson distribution tends to the normal, and indeed this is the case.

THEOREM 4.7. *The limiting form of the Poisson distribution as $r \to \infty$ is the standard form of the normal distribution.*

Proof. The characteristic function of the Poisson distribution is, from Eqns (3.26) and (4.43),

$$\phi_k(t) = \mathrm{e}^{-r} \exp(r\,\mathrm{e}^{it}).$$

If we now transform the distribution to standard measure by the relation

$$x = (k - \mu)/\sigma,$$

then

$$\begin{aligned}\phi_k(t) &= \int_{-\infty}^{\infty} dx\, f(x) \exp[it(\sigma x + \mu)], \\ &= \mathrm{e}^{it\mu}\phi_x(\sigma t).\end{aligned}$$

From Eqn (4.46) we have seen that

$$\mu = \sigma^2 = r,$$

and so

$$\phi_x(t) = \mathrm{e}^{-itr^{\frac{1}{2}}}\,\mathrm{e}^{-r} \exp[r\mathrm{e}^{itr^{-\frac{1}{2}}}],$$

and

$$\ln \phi_x(t) = -\,itr^{\frac{1}{2}} - r + r\exp(itr^{-\frac{1}{2}}).$$

If we now let $r \to \infty$, keeping t finite, we may expand the exponential and we find

$$\ln\phi_x(t) = -\,t^2/2 + O(r^{-\frac{1}{2}}).$$

Thus, for any finite t

$$\phi(t) \to \exp(-\,t^2/2).$$

which is the form of the c.f. of a standardized normal distribution [*cf.* Eqn (4.9)], and so, by the Inversion Theorem, the associated density function is the standardized form of the normal distribution.

The rate of convergence to normality is the same as for the binomial distribution [*cf.* Theorem 4.6], and so, in particular, the normal approximations to the Poisson distribution is quite adequate for values of $r \gtrsim 8$.

5 Sampling

5.1 Basic ideas

Since, in practice, we only have access to a sample of the whole population of events we have to consider carefully how best to choose a method of characterizing the sample such that our conclusions regarding the population remain relatively stable from one sample to the next. Clearly the measure selected should correspond to a parameter which varies little from sample to sample. In this section we will consider the desirable properties of such samples. Firstly, some definitions will be necessary.

DEFINITION 5.1. If $x_1, x_2, \ldots, x_n$ denotes a set of numerical values of n observations selected from a larger set then the set of values is called a *sample of size n.*

DEFINITION 5.2. A numerical value determined from some, or all, of the values of a sample is called a *statistic.*

Just as in Chapter 3 we defined certain useful population parameters we shall now define similar quantities to describe the corresponding sample statistic. It is conventional to use Greek letters for population parameters and Roman for sample parameters.

DEFINITION 5.3. The *sample mean* of a sample of size n is defined by

$$\bar{x}_n \equiv \bar{x} = \frac{1}{n} \sum_{i=1}^{n} x_i. \tag{5.1}$$

DEFINITION 5.4. The *sample variance* of a sample of size n is defined by

$$s^2 = \frac{1}{n-1} \sum_{i=1}^{n} (x_i - \bar{x})^2, \tag{5.2}$$

and, similarly, s is the sample standard deviation.

The sample variance has been defined in a way not exactly analogous to the definition of the population variance, Eqn (3.3).

Here we have replaced the factor $1/N$ of Eqn (3.3) by $1/(n-1)$. As will become clear in Chapter 7 this is to ensure that the expected value of all statistics of a given kind computed from samples of size n be equal to the corresponding population parameter.

We must also discuss how one can obtain the distribution of a sample statistic. A formal solution to this problem is as follows. Let $x_1, \ldots, x_n$ be a random sample of size n from a density $f(x)$. We wish to find the distribution of a sample statistic $y(x_1, \ldots, x_n)$. The distribution function of y is given by

$$F(y) = \int \cdots \int \prod_{i=1}^{n} f(x_i)\, dx_i,$$

where the integral is taken over the region such that $y \geqslant y(x_1, \ldots, x_n)$. In practice, it is often convenient to let $y(x_1, \ldots, x_n)$ be a new variable and then choose $n - 1$ other variables (functions of x_i) such that the n-dimensional integrand above takes a simple form. We will illustrate this by an example.

EXAMPLE 5.1. We shall find the sampling distribution of the mean $\bar{x}_n$ of a sample of size n drawn from the Cauchy distribution of Section 4.4, i.e.

$$f(x) = \frac{1}{\pi} \cdot \frac{1}{1 + x^2}, \qquad -\infty \leqslant x \leqslant \infty.$$

If we choose new variables $u_i = x_i$ $(i = 1, \ldots, n - 1)$ and let $u_n = \bar{x}_n$ then the Jacobian of the transformation is

$$J = \frac{\partial(x_1, x_2, \ldots, x_n)}{\partial(\bar{x}_n, u_1, u_2, \ldots, u_{n-1})},$$

and the distribution function becomes

$$F(\bar{x}_n) = \int \cdots \int f(x_1) \ldots f(x_n)\, J\, d\bar{x}_n' \prod_{i=1}^{n-1} dx_i,$$

taken over the region such that $\bar{x}_n' \geqslant \frac{1}{n} \sum x_i$. Thus

$$F(\bar{x}_n) = \int_{-\infty}^{\bar{x}_n} du_n \int_{-\infty}^{\infty} \cdots \int_{-\infty}^{\infty} \left(\frac{n}{\pi^n}\right) \prod_{j=1}^{n-1} \frac{du_j}{(1 + u_j^2)\left[1 + \left(nu_n - \sum_{i=1}^{n-1} u_i\right)^2\right]},$$

and the density function of $\bar{x}_n$ is given by the $(n - 1)$-fold integration in u_j $(j = 1, \ldots, n - 1)$. This integral can be evaluated but the algebra is rather lengthy. The result is

$$f(\bar{x}_n) = \frac{1}{\pi} \cdot \frac{1}{1 + \bar{x}_n^2}, \tag{5.3}$$

which is the same form as the population density.

Another useful method of finding sampling distributions is to find the m.g.f., or c.f., for the statistic, and then use either Theorem 3.2, or the Inversion Theorem to identify its density function. This technique is very practical and we shall have occasion to use it later.

5.2 Sampling distributions: theorems

In this subsection we will consider a number of definitions and theorems relating to sampling distributions in general. These will be useful when we consider more practical distributions in Chapter 6.

DEFINITION 5.5. Let S denote a sample of n observations x_i ($i = 1, 2, \ldots, n$) selected at random. The sample S is called a *random sample with replacement* (or a *simple random sample*) if, in general, the observation x_{n-1} is returned to the population before x_n is selected. If x_{n-1} is not so returned then S is called a *random sample without replacement.*

Sampling with replacement implies, of course, that it is indeed possible to return the "observation" to the population, as for example is the case when drawing cards from a deck. In most practical situations this is not possible and the sampling is without replacement.

The following theorems hold for S.

THEOREM 5.1. *Let N denote the size of any finite population and n the size of a sample without replacement, then for all possible samples of size n the mean of the means $\mu_{\bar{x}}$ is equal to the population mean μ, and the variance of the means ${\sigma_{\bar{x}}}^2$ is equal to the variance of the population σ^2 multiplied by* $(N-n)/[n(N-1)]$, i.e.

$$\mu_{\bar{x}} = \mu, \tag{5.4}$$

$${\sigma_{\bar{x}}}^2 = \frac{\sigma^2}{n}\left(\frac{N-n}{N-1}\right). \tag{5.5}$$

If the selection is with replacement then

$$\mu_{\bar{x}} = \mu, \tag{5.6}$$

$${\sigma_{\bar{x}}}^2 = \frac{\sigma^2}{n}. \tag{5.7}$$

If we consider, instead of a discrete finite population, a discrete but infinite one then it is clear that sampling with and without replacement lose their distinction and Eqns (5.6) and (5.7) hold. For continuous infinite populations we are led to the following theorem.

THEOREM 5.2. *Let x be a continuous random variable distributed with mean μ, variance σ^2 and density function $f(x)$. Let random samples of size n be drawn from this distribution. Then the sampling distribution of means has mean $\mu_{\bar{x}}$ equal to the population mean, and variance $\sigma_{\bar{x}}^2$ equal to the population variance σ^2 times a factor $1/n$, i.e.*

$$\mu_{\bar{x}} = \mu, \tag{5.8}$$

$$\sigma_{\bar{x}}^2 = \frac{\sigma^2}{n}. \tag{5.9}$$

As an example of the proofs of these theorems we shall prove Eqn (5.9).

Proof. By definition

$$\begin{aligned} \sigma_{\bar{x}}^2 &= E[(\bar{x} - E[\bar{x}])^2], \\ &= E[(\bar{x} - \mu)^2], \end{aligned}$$

which may be written

$$\sigma_{\bar{x}}^2 = \frac{1}{n^2} E\left[\left(\sum_{i=1}^{n} (x_i - \mu\right)^2\right]. \tag{5.10}$$

If we expand the square on the right hand side there are n terms of the form $(x_i - \mu)^2$ which gives contributions

$$\int_{-\infty}^{\infty} (x_i - \mu)^2 f(x_i)\, dx_i = \sigma^2. \tag{5.11}$$

The remaining terms are of the form $(x_i - \mu)(x_j - \mu)$ with $i < j$ and contribute terms

$$\begin{aligned} &\iint (x_i - \mu)(x_j - \mu) f(x_i) f(x_j)\, dx_i\, dx_j \\ &= \int_{-\infty}^{\infty} (x_i - \mu) f(x_i)\, dx_i \int_{-\infty}^{\infty} (x_j - \mu) f(x_j)\, dx_j \\ &= 0. \end{aligned} \tag{5.12}$$

Thus (5.10) becomes

$$\sigma_{\bar{x}}^2 = \frac{1}{n^2}\sum_{i=1}^{n} \sigma^2 = \frac{\sigma^2}{n}. \tag{5.13}$$

The above theorems are of considerable importance because they show that as the sample size increases the variance of the sample mean decreases, and thus the probability that the sample mean is a good estimation of the population mean increases. This result may be stated formally as the Weak Law of Large Numbers.

Weak law of large numbers

Let x_i be a population of independent random variables with mean μ and finite variance. Let $\bar{x}_n$ be the mean of a sample of size n

$$\bar{x}_n = \frac{1}{n}\sum_{i=1}^{n} x_i.$$

Then, given any $\varepsilon > 0$ and δ in the range of $0 < \delta < 1$, there exists an integer n such that for all $m \geqslant n$

$$P[|\bar{x}_m - \mu| < \varepsilon] \geqslant 1 - \delta. \tag{5.14}$$

The Weak Law of Large Numbers tell us that $|\bar{x}_n - \mu|$ will ultimately be very small but does not exclude the possibility that for some finite n it could be large. Since, in practice, we can only have access to finite samples this possibility could be of some importance. Fortunately there exists the so-called *Strong Law of Large Numbers*, which, in effect, states that the probability of such an occurrence is extremely small. It is the Laws of Large Numbers which ensure that the empirical definition of probability we have adopted concurs in practice with the axiomatic one.

The Weak Law of Large Numbers is a special case of Tchebysheff's Inequality which may be stated as follows:

Tchebysheff's inequality

Let $f(x)$ be a density function with mean μ and finite variance σ^2. Let p be any positive number, and let $\bar{x}_n$ be the mean of a random sample of size n drawn from $f(x)$. Then

$$P\left[|\bar{x}_n - \mu| \leqslant \frac{p\sigma}{n^{\frac{1}{2}}}\right] \geqslant 1 - \frac{1}{p^2}. \tag{5.15}$$

Proof. Let the density function of $\bar{x}_n$ be $g(\bar{x}_n)$. Then from Theorem 5.2 we have

$$\sigma_{\bar{x}}^2 = \frac{\sigma^2}{n} = \int_{-\infty}^{\infty} (\bar{x}_n - \mu)^2 g(\bar{x}_n)\, d\bar{x}_n, \tag{5.16}$$

$$= \int_{-\infty}^{\mu-(p\sigma/n^{\frac{1}{2}})} (\bar{x}_n - \mu)^2 g(\bar{x}_n)\, d\bar{x}_n + \int_{\mu-(p\sigma/n^{\frac{1}{2}})}^{\mu+(p\sigma/n^{\frac{1}{2}})} (\bar{x}_n - \mu)^2 g(\bar{x}_n)\, d\bar{x}_n$$

$$+ \int_{\mu+(p\sigma/n^{\frac{1}{2}})}^{\infty} (\bar{x}_n - \mu)^2 g(\bar{x}_n)\, d\bar{x}_n. \tag{5.17}$$

Now if we replace $(\bar{x}_n - \mu)^2$ by $p^2\sigma^2/n$ in the first integral the value of this integral will clearly not decrease. The same argument holds for the third integral, and the second integral is nonnegative, so we may obtain from (5.17) the inequality

$$\frac{\sigma^2}{n} \geqslant \frac{p^2\sigma^2}{n}\left\{\int_{-\infty}^{\mu-(p\sigma/n^{\frac{1}{2}})} g(\bar{x}_n)\, d\bar{x}_n + \int_{\mu+(p\sigma/n^{\frac{1}{2}})}^{\infty} g(\bar{x}_n)\, d\bar{x}_n\right\}, \tag{5.18}$$

which, from the definition of a distribution function, is just the statement that

$$P\left[\,|\,\bar{x}_n - \mu\,| \geqslant \frac{p\sigma}{n^{\frac{1}{2}}}\right] \leqslant \frac{1}{p^2},$$

or

$$P\left[\,|\,\bar{x}_n - \mu\,| \leqslant \frac{p\sigma}{n^{\frac{1}{2}}}\right] \geqslant 1 - \frac{1}{p^2}, \tag{5.19}$$

which completes the proof.

The Weak Law of Large Numbers may be proved from Tchebysheff's inequality provided the population distribution has a finite variance. We merely choose $p = \delta^{-\frac{1}{2}}$ and $n > \sigma^2/\delta\varepsilon^2$, and substitute in Eqn (5.19). The bound given by (5.15) is usually weak, but if we restrict ourselves to the sampling distribution of the mean then we can state the most important theorem in statistics, the Central Limit Theorem.

THEOREM 5.3 (Central Limit Theorem). *Let the independent random variables x_i, of unknown density function, be identically distributed with mean μ and variance σ^2, both of which are finite. Then the distribution of the sample mean $\bar{x}_n$ tends to the normal distribution with mean μ and variance σ^2/n when* *n becomes large. Thus, if $u(t)$ is the standard form of the normal density function, then for arbitrary t_1 and t_2*

$$\lim_{n\to\infty} P\left\{t_1 \leqslant \frac{\bar{x}_n - \mu}{\sigma/n^{\frac{1}{2}}} \leqslant t_2\right\} = \int_{t_1}^{t_2} u(t)\,dt. \tag{5.20}$$

Proof. By applying the results on expected values given previously in Eqns (3.21) and (3.22) to m.g.f.'s it follows immediately that if the components of the sample are independent then the mean and variance of their sum $S = \sum_{i=1}^{n} x_i$ are given by

$$\mu_s = n\mu \quad \text{and} \quad \sigma_s^2 = n\sigma^2.$$

Now consider the variable

$$u = \frac{S - \mu_s}{\sigma_s} = \frac{1}{\sqrt{n}\sigma} \sum_{i=1}^{n} (x_i - \mu), \tag{5.21}$$

with c.f. $\phi_u(t)$. If $\phi_i(t)$ is the c.f. of $(x_i - \mu)$, then

$$\phi_u(t) = \prod_{i=1}^{n} \phi_i\left(\frac{t}{\sqrt{n}\sigma}\right),$$

but all the $(x_i - \mu)$ have the same distribution and so

$$\phi_u(t) = \left[\phi_i\left(\frac{t}{\sqrt{n}\sigma}\right)\right]^n. \tag{5.22}$$

Just as the m.g.f. can be expanded into an infinite series of moments so we can expand the c.f. Thus

$$\phi(t) = 1 + \sum_{r=1}^{\infty} \mu_r' \frac{(it)^r}{r!}, \tag{5.23}$$

and since the first two moments of $(x_i - \mu)$ are zero and σ^2 we have from (5.22) and (5.23)

$$\phi_u(t) = \left[1 - \frac{t^2}{2n} + O\left(\frac{1}{n}\right)\right]^n.$$

Thus, for fixed t, as $n \to \infty$

$$\phi_u(t) \to \mathrm{e}^{-t^2/2}, \tag{5.24}$$

which is the c.f. of a standardized normal distribution. Thus by the Inversion Theorem S is distributed as $N(S; \mu_s, \sigma_s^2)$ and hence $\bar{x}_n$ is distributed as $N(\bar{x}_n; \mu, \sigma^2/n)$.

The form of the Central Limit Theorem above is not the most general that can be given. For example, provided certain (weak) conditions on the third moments are obeyed then the condition that the x_i all have the same distribution can be relaxed, and it is possible to prove that the sampling distribution of *any* linear combination of independent random variables having arbitrary distributions but with finite means and variances tends to normality for large samples. There are also circumstances under which the assumption of independence can be relaxed.

The Central Limit Theorem applies to both discrete and continuous distributions and is a most remarkable theorem because nothing is said about the original density function, except that it have finite mean and variance, which in practice are seldom restrictions. However, this condition is essential. Thus we have seen in Example 5.1 that for the Cauchy distribution the distribution of $\bar{x}_n$ is

$$f(\bar{x}_n) = \frac{1}{\pi} \cdot \frac{1}{1 + \bar{x}_n^2},$$

i.e. the same as for a single observation. The failure of the Central Limit Theorem in this case can be traced to the infinite variance of the Cauchy distribution. It is the Central Limit Theorem which gives the normal distribution such a prominent position both theoretically and in practice. In particular, it allows (approximate) quantitative probability statements to be made in experimental situations where the exact form of the underlying distribution is unknown.

Just as we have been considering the sampling distribution of means we can also consider the sampling distribution of sums $T = \Sigma x_i$ of random samples of size n. If the random variable x is distributed with mean μ, and variance σ^2, then the sampling distribution of T has mean

$$\mu_T = n\mu, \tag{5.25}$$

and variance

$$\sigma_T^2 = \begin{cases} n\sigma^2 \left(\dfrac{N - n}{N - 1}\right) & \text{for sampling from a finite population of size } N \text{ without replacement} \\ n\sigma^2 & \text{otherwise.} \end{cases} \tag{5.26}$$

We shall conclude this section with a few results on the properties of linear combinations of means, since up to now we have been concerned mainly with sampling distributions of a single sample mean.

THEOREM 5.4. *Let*

$$l = \sum_{i=1}^{n} a_i x_i, \tag{5.27}$$

where a_i are real constants, and the x_i are random variables with mean μ_i, variances σ_i^2 and covariances σ_{ij} $(i, j = 1, 2, \ldots, n; i \neq j)$. Then

$$\mu_l = \sum_{i=1}^{n} a_i \mu_i, \tag{5.28}$$

and

$$\sigma_l^2 = \sum_{i=1}^{n} a_i^2 \sigma_i^2 + 2 \sum_{i<j} a_i a_j \sigma_{ij}, \tag{5.29}$$

$$= \sum_{i=1}^{n} a_i^2 \sigma_i^2, \text{ if the } x_i\text{'s are mutually independent.} \tag{5.30}$$

Proof. Let $f(x_1, x_2, \ldots, x_n)$ be the joint density function of $x_1, x_2, \ldots, x_n$. Then using Eqns (3.33)

$$\mu_l = \int_{-\infty}^{\infty} \cdots \int_{-\infty}^{\infty} l f(x_1, x_2, \ldots, x_n) \prod_{i=1}^{n} dx_i,$$

$$= \int_{-\infty}^{\infty} \cdots \int_{-\infty}^{\infty} \left(\sum_{j=1}^{n} a_j x_j \right) f(x_1, x_2, \ldots, x_n) \prod_{i=1}^{n} dx_i,$$

$$= \sum_{j=1}^{n} a_j \int_{-\infty}^{\infty} \cdots \int_{-\infty}^{\infty} x_j f(x_1, x_2, \ldots, x_n) \prod_{i=1}^{n} dx_i,$$

$$= \sum_{j=1}^{n} a_j \mu_j.$$

Also, using Eqns (3.34) and (3.36)

$$\sigma_l^2 = \int_{-\infty}^{\infty} \cdots \int_{-\infty}^{\infty} (l - \mu_l)^2 f(x_1, x_2, \ldots, x_n) \prod_{i=1}^{n} dx_i$$

$$= \int_{-\infty}^{\infty} \cdots \int_{-\infty}^{\infty} \left[\sum_{j=1}^{n} a_j (x_j - \mu_j) \right]^2 f(x_1, x_2, \ldots, x_n) \prod_{i=1}^{n} dx_i$$

$$= \sum_{j=1}^{n} a_j^2 \int_{-\infty}^{\infty} \cdots \int_{-\infty}^{\infty} (x_j - \mu_j)^2 f(x_1, x_2, \ldots, x_n) \prod_{i=1}^{n} dx_i$$

$$+ 2 \sum_{k<j} a_k a_j \int_{-\infty}^{\infty} \cdots \int_{-\infty}^{\infty} (x_k - \mu_k)(x_j - \mu_j) f(x_1, x_2, \ldots, x_n) \prod_{i=1}^{n} dx_i$$

$$= \sum_{i=1}^{n} a_i^2 \sigma_i^2 + 2 \sum_{k<j} a_k a_j \sigma_{kj}.$$

which completes the proof.

A useful corollary to the above theorem is as follows. Let $\bar{x}_i$ $(i = 1,2)$ be the mean of a random sample of size n_i drawn from an infinite population with mean μ_i and variance σ_i. If $\bar{x}_1$ and $\bar{x}_2$ are independently distributed, then

$$\mu_{\bar{x}_1 + \bar{x}_2} = \mu_1 \pm \mu_2, \tag{5.31}$$

and

$$\sigma_{\bar{x}_1 + \bar{x}_2}^2 = \sum_{i=1}^{2} \left(\frac{\sigma_i^2}{n_i} \right). \tag{5.32}$$

These results follow immediately from Eqns (5.28) and (5.29), and Theorem 5.2, by the substitution $x_1 = \bar{x}_1$ and $x_2 = \bar{x}_2$ with $a_1 = a_2 = 1$ for the first case and $a_1 = -a_2 = 1$ for the second.

5.3 Experimental errors and their propagation

In the preceding sections we have been concerned with theoretical statistics only. In this subsection we will provide the link between theoretical statistics and experimental situations.

In an experimental observation one can never measure the value of a quantity with absolute precision, that is one can never reduce the error on the measurement to zero. The *precision* of a measurement will be taken to mean the smallness of the error. By *accuracy* we shall mean the deviation of the observation from the "true" value, assuming that such a concept is meaningful. Thus there may exist, in addition to fluctuations in the measurement process which limit the precision, unknown systematic errors which limit the accuracy, In general, the only errors that we can deal with here are the former type, and the conventional measure of this type of error is taken to be the standard deviation σ. In this case, it is called the *standard error.* This definition of the error is, of course, arbitrary, and some workers still use the older *probable error* p which is defined by

$$\int_{\mu-p}^{\mu+p} f(x)\, dx = \tfrac{1}{2}.$$

Unfortunately some authors do not say which form of error they are using, or even multiply their errors by an arbitrary factor "to be on the safe side" when quoting results. Needless to say such practices render statistical analyses meaningless and are to be discouraged.

Consider, for example, an idealized nuclear counting experiment for a scattering process. The number of trials, i.e. the number of incoming particles, is very large, but the probability of a scatter, p is very small. In this situation the Poisson distribution (Section 4.7) is applicable, and as we have seen in Eqn (4.46), if $N_e = np$ is the total number of counts recorded then $\sigma = (N_e)^{\frac{1}{2}}$. The result of the experiment would be given as

$$N = N_e \pm \Delta N, \tag{5.33}$$

where

$$\Delta N = (N_e)^{\frac{1}{2}}. \tag{5.34}$$

If the population distribution is unknown then we can consider the sampling distribution. For example, from a set of observations x_i we know that an estimate of the mean is the sample mean

$$\bar{x} = \frac{1}{n} \sum_{i=1}^{n} x_i, \tag{5.35}$$

and the Laws of Large Numbers ensure that $\bar{x}$ is a good estimate for large n. The variance of $\bar{x}$ is

$$\sigma_{\bar{x}}^2 = \frac{\sigma^2}{n}, \tag{5.36}$$

so to calculate $\sigma_{\bar{x}}^2$ we need to estimate σ^2. We have seen that the sample variance is

$$s^2 = \frac{1}{n-1} \sum_{i=1}^{n} (x_i - \bar{x})^2, \tag{5.37}$$

and thus

$$\sigma_{\bar{x}}^2 = \frac{1}{n(n-1)} \sum_{i=1}^{n} (x_i - \bar{x})^2. \tag{5.38}$$

The experimental result would then be quoted as

$$x = \bar{x}_e \pm \Delta x, \tag{5.39}$$

where

$$\Delta x = \sigma_{\bar{x}} = \left[\frac{1}{n(n-1)} \sum_{i=1}^{n} (x_i - \bar{x})^2\right]^{1/2}. \tag{5.40}$$

Now by the Central Limit Theorem we know that the distribution of the sample means is approximately normal, and so Eqn (5.39) may be interpreted (*cf.* Section 4.2) as

$$P[\bar{x}_e - \Delta x \leqslant x \leqslant \bar{x}_e + \Delta x] \simeq 68{\cdot}3\,\%$$
$$P[\bar{x}_e - 2\Delta x \leqslant x \leqslant \bar{x}_e + 2\Delta x] \simeq 95{\cdot}4\,\%$$
$$P[\bar{x}_e - 3\Delta x \leqslant x \leqslant \bar{x}_e + 3\Delta x] \simeq 99{\cdot}7\,\%.$$

Thus, even though the form of the underlying distribution of x is unknown, the Central Limit Theorem has enabled us to make an approximate quantitative statement about the probability of the true value of x lying within a specified range.

5.3.1. Propagation of Errors

If we have a function y of the p parameters θ_i $(i = 1, \ldots, p)$ i.e.

$$y \equiv y(\boldsymbol{\theta}) = y(\theta_1, \theta_2, \ldots, \theta_p),$$

then we are often interested in knowing the approximate error on y, given that we know the errors on θ_i. If the true values of θ_i are θ_i^* (in practice estimates of these quantities would usually have to be used), and the quantities $(\theta_i - \theta_i^*)$ are small, then a Taylor expansion of $y(\boldsymbol{\theta})$ about the point $\boldsymbol{\theta} = \boldsymbol{\theta}^*$ gives, to first order

$$y(\boldsymbol{\theta}) = y(\boldsymbol{\theta}^*) + \sum_{i=1}^{p} (\theta_i - \theta_i^*) \left.\frac{\partial y(\boldsymbol{\theta})}{\partial \theta_i}\right|_{\boldsymbol{\theta}=\boldsymbol{\theta}^*}, \tag{5.41}$$

Now

$$\begin{aligned} \operatorname{var} y(\boldsymbol{\theta}) &= E[(y(\boldsymbol{\theta}) - E[y(\boldsymbol{\theta})])^2] \\ &\simeq E[(y(\boldsymbol{\theta}) - y(\boldsymbol{\theta}^*))^2]. \end{aligned} \tag{5.42}$$

Using (5.41) in (5.42) gives

$$\operatorname{var} y(\boldsymbol{\theta}) \simeq \sum_{i=1}^{p} \sum_{j=1}^{p} \left.\frac{\partial y(\boldsymbol{\theta})}{\partial \theta_i}\right|_{\boldsymbol{\theta}=\boldsymbol{\theta}^*} \left.\frac{\partial y(\boldsymbol{\theta})}{\partial \theta_j}\right|_{\boldsymbol{\theta}=\boldsymbol{\theta}^*} E[(\theta_i - \theta_i^*)(\theta_j - \theta_j^*)]. \tag{5.43}$$

Now, from Eqn (3.36) we see that

$$V_{ij} = E[(\theta_i - \theta_i^*)(\theta_j - \theta_j^*)],$$

is the matrix of variances and covariances of the parameters θ_i (the *variance matrix*). Thus, if we set

$$(\Delta y)^2 = \text{var}\, y,$$

we have

$$(\Delta y)^2 = \sum_{i=1}^{p} \sum_{j=1}^{p} \left\{ \frac{\partial y(\boldsymbol{\theta})}{\partial \theta_i}\bigg|_{\boldsymbol{\theta}=\boldsymbol{\theta}^*} V_{ij} \frac{\partial y(\boldsymbol{\theta})}{\partial \theta_j}\bigg|_{\boldsymbol{\theta}=\boldsymbol{\theta}^*} \right\}. \tag{5.44}$$

Equation (5.44) is often referred to as the *Law of Propagation of Errors.* If the errors are uncorrelated (i.e. cov $(\theta_i, \theta_j) = 0$) then

$$V_{ij} = \begin{cases} (\Delta\theta_i)^2, & i = j \\ 0, & i \neq j, \end{cases}$$

and (5.44) reduces to

$$(\Delta y)^2 = \sum_{i=1}^{p} \left[\frac{\partial y(\boldsymbol{\theta})}{\partial \theta_i}\bigg|_{\boldsymbol{\theta}=\boldsymbol{\theta}^*} \Delta\theta_i \right]^2. \tag{5.45}$$

When using these expressions one should always ensure that the quantities $\Delta\theta_i \equiv \theta_i - \theta_i^*$ are small enough to justify truncation of the Taylor series (5.41) at the first order in $\Delta\theta_i$.

6 Sampling Distributions Associated with the Normal

The special position held by the normal distribution, mainly by virtue of the Central Limit Theorem, is reflected in the prominent positions of certain related distributions. In this chapter we will consider the basic properties of three frequently used distributions; the Chi-square, the Student t and F distributions. These arise when sampling from normal populations, and are widely used in estimation problems (i.e. finding the best values of parameters), and in the testing of hypotheses. These topics will be discussed in detail in Chapters 7–11.

6.1 Chi-square distribution

If we wish to concentrate on a measure to describe the dispersion of a population then we consider the sample variance. The chi-square distribution is introduced for problems involving this quantity.

THEOREM 6.1. *If x_i $(i = 1, 2, \ldots, \nu)$ is a sample of ν random variables normally and independently distributed with means μ_i and variances σ_i^2, then the statistic*

$$\chi^2 = \sum_{i=1}^{\nu} \left[\frac{x_i - \mu_i}{\sigma_i}\right]^2 \tag{6.1}$$

is distributed with density function

$$f(\chi^2; \nu) = \frac{1}{2^{\nu/2}\Gamma(\nu/2)} \chi^{2[(\nu/2)-1]} \exp(-\chi^2/2), \quad \chi^2 > 0, \tag{6.2}$$

mean ν and variance 2ν.

This is known as the χ^2-distribution (chi-square) with ν *degrees of freedom.* (It is perhaps unfortunate that the symbol χ^2 should have been chosen for both the statistic and the distribution). The symbol $\Gamma(x)$ in Eqn (6.2) is the Gamma function, defined by the integral

$$\Gamma(x) = \int_0^\infty du\, e^{-u} u^{x-1}, \qquad 0 < x < \infty \tag{6.3}$$

and is frequently encountered in sampling distributions associated with the normal distribution. To prove this theorem we shall again use the method of characteristic functions.

Proof. Let

$$\chi^2 = \sum_{i=1}^{\nu} \left(\frac{x_i - \mu_i}{\sigma_i} \right)^2 = \sum_{i=1}^{\nu} z_i^2.$$

Then the z_i are distributed as $N(z_i; 0, 1)$, and from Example 3.2 we know that the quantities $u_i \equiv z_i^2$ have density functions

$$\frac{1}{(2\pi u_i)^{1/2}} \exp(-u_i/2).$$

Thus the c.f. of u_i is

$$\begin{aligned} \phi_i(t) &= \int_0^\infty du_i \frac{1}{(2\pi u_i)^{1/2}} \, e^{-u_i/2} \, e^{itu_i} \\ &= (1 - 2it)^{-1/2}, \qquad (u_i \geqslant 0). \end{aligned}$$

Now since the random variables u_i are independently distributed, if $\phi(t)$ is the c.f. of χ^2, then

$$\phi(t) = \sum_{i=1}^{\nu} \phi_i(t) = (1 - 2it)^{-\nu/2}. \tag{6.4}$$

The density function of χ^2 is now obtainable from the Inversion Theorem. It is

$$f(\chi^2; \nu) = \frac{1}{2\pi} \int_0^\infty dt \, (1 - 2it)^{-\nu/2} \, e^{-i\chi^2 t}$$

This integral may be evaluated using the definition of $\Gamma(x)$ in Eqn (6.3) and gives

$$f(\chi^2; \nu) = \frac{1}{2^{\nu/2} \, \Gamma(\nu/2)} \chi^{2[(\nu/2)-1]} \exp(-\chi^2/2),$$

which is the required result.

The m.g.f. is obtainable directly from Eqns (6.4) and (3.23), and is

$$M(t) = (1 - 2t)^{-\nu/2}.$$

It follows that

$$\mu = \nu \quad \text{and} \quad \sigma^2 = 2\nu,$$

which completes the proof.

Graphs of $f(\chi^2; \nu)$ and its distribution function $F(\chi^2; \nu)$ for $\nu = 1, 4$ and 10 are shown in Fig. (6.1), and tables of $F(\chi^2; \nu)$ are given in Appendix D.

The third and fourth moments about the mean may also be found from the m.g.f. $M(t)$. They are

$$\mu_3 = 8\nu; \quad \mu_4 = 12\nu(\nu + 4),$$

giving

$$\beta_1 = \frac{8}{\nu}; \quad \beta_2 = 3\left(1 + \frac{4}{\nu}\right),$$

which tend to the values for the normal distribution as $\nu \to \infty$. The χ^2 distribution does indeed tend to normality for large samples and we can demonstrate this by constructing the c.f. for the standardized variable

$$y \equiv \left(\frac{\chi^2 - \mu}{\sigma}\right) = \left(\frac{\chi^2 - \nu}{(2\nu)^{1/2}}\right).$$

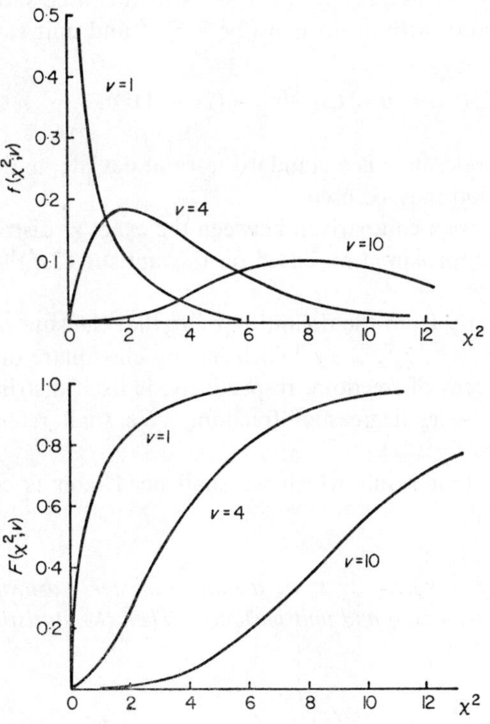

FIG. 6.1. The chi-square density $f(\chi^2; \nu)$, and its distribution function $F(\chi^2; \nu)$, for $\nu=1$, 4 and 10.

From (6.4) it is

$$\phi_y(t) = \exp\left[\frac{-ivt}{(2v)^{1/2}}\right]\left[1 - \frac{2it}{(2v)^{1/2}}\right]^{-v/2},$$

and hence, taking logarithms and letting $v \to \infty$ gives

$$\ln \phi_y(t) \to \frac{-ivt}{(2v)^{1/2}} - \frac{v}{2}\left[\frac{-2it}{(2v)^{1/2}} - \tfrac{1}{2}\left(\frac{2it}{(2v)^{1/2}}\right)^2 \cdots\right], \qquad (v \to \infty),$$

i.e.

$$\phi_y(t) \to \exp(-t^2/2),$$

which is the c.f. of the standardized normal distribution. Thus, by the Inversion Theorem, the χ^2 distribution tends to normality as $v \to \infty$, although the rate of convergence is rather slow.

Because the χ^2 distribution is a one-parameter family it frequently happens that tabulated values do not exist for precisely the range one requires. In such cases a very useful statistic is $(2\chi^2)^{1/2}$ which can be shown to tend very rapidly to normality with mean $\mu = (2v - 1)^{1/2}$ and unit variance. Thus the statistic

$$u = (2\chi^2)^{1/2} - (2v - 1)^{1/2},$$

for even quite moderate v is a standard normal deviate, and so tables of the normal distribution may be used.

Table (6.1) shows a comparison between the exact χ^2 distribution function and the normal approximation based on the statistic $(2\chi^2)^{1/2}$ for a range of values, of v and χ^2.

It follows directly from the definition of χ^2 that the sum of n independent random variables $\chi_1^2, \chi_2^2, \ldots, \chi_n^2$ each having chi-square distributions with $v_1, v_2, \ldots, v_n$ degrees of freedom, respectively, is itself distributed as χ^2 with $v = v_1 + v_2 + \ldots + v_n$ degrees of freedom. We shall refer to this as the *additive property of* χ^2.

Another important result which we shall need later is contained in the following theorem.

THEOREM 6.2. *Let $x_1, x_2, \ldots, x_v$ be a sample of size v drawn from a normal population with mean zero and unit variance. Then the statistic*

$$u = \sum_{i=1}^{v} (x_i - \bar{x})^2,$$

s distributed as χ^2 with $(v - 1)$ degrees of freedom.

TABLE 6.1 Values of $P[\chi^2 \geqslant \chi_\alpha^2]$ for $\nu = 5, 10$, and 20 and $\chi_\alpha^2 = 2, 5, 10, 20, 30$ using the exact χ^2 distribution function, and the normal approximation.

	ν					
	5		10		20	
χ_α^2	exact	approx.	exact	approx.	exact	approx.
2	0·849	0·841	0·996	0·991		
5	0·416	0·436	0·891	0·885		
10	0·075	0·071	0·441	0·456	0·968	0·963
20	0·001	0·001	0·029	0·024	0·458	0·462
30			0·001	0·000	0·070	0·067

Proof. Consider the transformation of variables defined by

$$u_1 = (x_1 - x_2)/\sqrt{2},$$

$$u_2 = (x_1 + x_2 - 2x_3)/\sqrt{6},$$

....

$$u_{\nu-1} = (x_1 + x_2 + \ldots + x_{\nu-1} - (\nu - 1)\, x_\nu)/\sqrt{[\nu(\nu-1)]},$$

$$u_\nu = (x_1 + x_2 + \ldots + x_\nu)/\sqrt{\nu}.$$

It can be easily verified that if the x_i are independently normally distributed with mean zero and unit variance then so are the variables u_i. Now consider the statistic

$$\sum_{i=1}^{\nu} (x_i - \bar{x})^2 = \sum_{i=1}^{\nu} (x_i^2) - \nu\bar{x}^2$$

$$= \sum_{i=1}^{\nu} u_i^2 - u_\nu^2 = \sum_{i=1}^{\nu-1} u_i^2.$$

Thus the sum of squares of v standard normal variates measured from their mean is distributed as the sum of $(v - 1)$ normal variates with mean zero. It follows from Theorem 6.1 that the statistic

$$u = \sum_{i=1}^{v} (x_i - \bar{x})^2,$$

is distributed as χ^2 with $(v - 1)$ degrees of freedom.

In general, if the parent population has variance σ^2, then

$$\chi^2 = \frac{1}{\sigma^2} \sum_{i=1}^{v} (x_i - \bar{x})^2,$$

is distributed as χ^2 with $(v - 1)$ degrees of freedom. Moreover, since the sample variance is

$$s^2 = \frac{\sigma^2 \chi^2}{v - 1},$$

it follows that $(v - 1)s^2/\sigma^2$ is distributed as χ^2 with $(v - 1)$ degrees of freedom, *independent* of the sample mean $\bar{x}$. Thus we have also proved the following theorem.

THEOREM 6.3. *The sample mean and sample variance are independent random variables when sampling randomly from normal populations.*

This somewhat surprising result is very important in practice and we will use it later to construct the sampling distribution known as the Student t distribution.

If we assume that a sample is drawn at random from a single normal population with mean μ and variance σ^2 then from (6.1)

$$\chi^2 = \frac{1}{\sigma^2} \sum_{i=1}^{v} (x_i - \mu)^2.$$

However, since the mean of the population is rarely known, in these cases it is more useful to use the results of Theorem 6.2. In that case χ^2 is distributed with $(v - 1)$ degrees of freedom if $\bar{x}$ is used instead of μ. *In general, the number of degrees of freedom must be reduced by one for each parameter estimated from the data.*

The χ^2 distribution is one of the most important sampling distributions occurring in physics. The density function is a family of curves, but a useful table may be constructed by calculating the proportion α of the area under the χ^2 curves to the *right* of the point χ_α^2, i.e. points such that

$$P[\chi^2 \geqslant \chi_\alpha^2] = \alpha = \int_{\chi_\alpha^2}^{\infty} f(\chi^2; \nu)\, d\chi^2.$$

Such points are called *percentage points* of the χ^2-distribution and may be deduced from the tables in Appendix D. They are also shown graphically in Fig. (6.2). A point of interest about these curves is that for a fixed value of P, $\chi^2/\nu \to 1$ as $\nu \to \infty$.

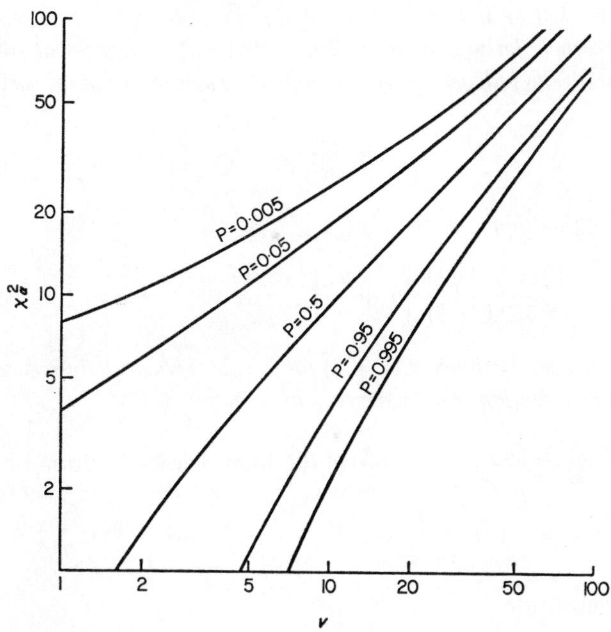

FIG. 6.2. Percentage points of the chi-square distribution. $P = P[\chi^2 \geqslant \chi_\alpha^2]$.

6.2 Student t distribution

The Central Limit Theorem told us that the distribution of the sample mean $\bar{x}$ was approximately normal with mean μ (the population mean) and variance σ^2/n (where σ^2 is the population variance and n is the sample size). Thus, in standard measure, the statistic

$$u = \left(\frac{\bar{x} - \mu}{\sigma/\sqrt{n}}\right),$$

is approximately normally distributed with mean 0 and unit variance for large n. However, in experimental situations neither the mean nor the population variance are known, and must be replaced by estimates calculated from the sample. One can safely replace σ^2 by the sample variance s^2 for large $n \gtrsim 50$, but for small n the statistic will not be approximately normally distributed and serious loss of meaning in interpretation will occur. In such cases the student t distribution must be used. This distribution, which we will discuss in this section, enables one to use the sample variance, as well as the sample mean, to make statements about the population mean. We shall concentrate our results in four important theorems.

THEOREM 6.4. *Let u have a normal distribution with mean zero and unit variance. Further, let w have a χ^2 distribution with v degrees of freedom, and let u and $(w)^{1/2}$ be independently distributed. Then the random variable*

$$t = \frac{u}{(w/v)^{1/2}},$$

has a density function

$$f(t; v) = \frac{\Gamma[(v+1)/2]}{(\pi v)^{1/2}\Gamma(v/2)}\left[1 + \frac{t^2}{v}\right]^{-(v+1)/2}, \qquad -\infty < t < \infty, \tag{6.5}$$

with mean zero and variance $v/(v-2)$ for $v > 2$. The statistic t is said to have a Student t distribution, with v degrees of freedom.

Proof. From Eqns (6.2) and (4.10) the joint density function of u and w is

$$f(u, w; v) = \frac{1}{(2\pi)^{1/2}}\,\mathrm{e}^{-u^2/2}\,\frac{1}{\Gamma(v/2)2^{v/2}}\,w^{(v-2)/2}\,\mathrm{e}^{-w/2}. \tag{6.6}$$

If we now substitute

$$u = t\left(\frac{w}{v}\right)^{1/2},$$

then (6.6) becomes

$$f(t, w; v) = \frac{\mathrm{e}^{-t^2w/2v}\,\mathrm{e}^{-w/2}\,w^{(v-2)/2}}{(2\pi)^{1/2}\Gamma(v/2)2^{v/2}}, \tag{6.7}$$

and

$$f(t; v) = \int_0^\infty f(t, w; v)\,dw.$$

This integral may be evaluated directly using Eqn (6.3) and gives Eqn (6.5).

To find the mean and variance we again use the method of the m.g.f. From (6.5) and (3.23) we see that moments of order r exist only for $r < \nu$, and are zero by symmetry for odd order moments. For even order moments direct integration gives

$$\mu_{2r} = \nu^r \frac{\Gamma(r + 1/2)\Gamma(\nu/2 - r)}{\Gamma(1/2)\Gamma(\nu/2)}, \qquad 2r < \nu. \tag{6.8}$$

The mean and variance may be obtained from (6.8). They are

$$\mu = 0; \qquad \sigma^2 = \frac{\nu}{\nu - 2}.$$

The second theorem specifies the distribution of the difference of the sample mean and the population mean with respect to the sample variance.

THEOREM 6.5. *Let x_i $(i = 1, \ldots, n)$ be a random sample of size n drawn from a normal population with mean μ and variance σ^2. Then the statistic*

$$t = \frac{\sqrt{n}}{s}(\bar{x} - \mu), \tag{6.9}$$

where

$$s^2 = \frac{1}{n-1}\sum_{i=1}^{n}(x_i - \bar{x})^2, \tag{6.10}$$

and

$$\bar{x} = \frac{1}{n}\sum_{i=1}^{n} x_i,$$

is distributed as the Student t distribution with $(n - 1)$ degrees of freedom.

Proof. If the mean and variance of the population are μ and σ^2, respectively, then the statistic

$$u = \left(\frac{\bar{x} - \mu}{\sigma/\sqrt{n}}\right),$$

is distributed as $N(u; 0, 1)$. Furthermore, from Theorem 6.2 we know that the statistic

$$w = (n - 1)\frac{s^2}{\sigma^2},$$

is distributed as χ^2 with $n - 1$ degrees of freedom. Therefore, from Theorem 6.4 the statistic

$$t = \frac{u}{[w/(n-1)]^{1/2}} = \frac{\sqrt{n}}{s}(\bar{x} - \mu),$$

is distributed as t with $(n - 1)$ degrees of freedom.

The third theorem concerns the asymptotic behaviour of the t distribution.

THEOREM 6.6. *As the number of degrees of freedom of the Student t distribution approaches infinity, the distribution tends to the normal distribution in standard form.*

Proof. If in Eqn (6.8) we use Stirling's approximation in the gamma functions, i.e.

$$\Gamma(\nu + 1) \to (2\pi)^{1/2}\, \nu^{\nu + 1/2}\, e^{-\nu}, \qquad (\nu \to \infty), \tag{6.11}$$

then

$$\mu_{2r} \to \frac{(2r)!}{2^r r!}. \tag{6.12}$$

However, from (4.7) we know that this is the expression for the moments of the normal distribution expressed in standard measure. Thus, by Theorem 3.2, the Student t distribution tends to a normal distribution with mean zero and unit variance.

The final theorem concerns the t distribution when two normal populations are involved.

THEOREM 6.7. *Let random samples $x_{11}, x_{12}, \ldots, x_{1n_1}$, and $x_{21}, x_{22}, \ldots, x_{2n_2}$ of sizes n_1 and n_2, respectively, be independently drawn from two normal populations* 1 *and* 2 *with means μ_1 and μ_2, and the same variance σ^2. Then, if we define*

$$\bar{x}_i = \frac{1}{n_i}\sum_{j=1}^{n_i} x_{ij}, \qquad i = 1, 2,$$

the statistic

$$t = \frac{(\bar{x}_1 - \bar{x}_2) - (\mu_1 - \mu_2)}{\left[S_p^{\,2}\left(\dfrac{1}{n_1} + \dfrac{1}{n_2}\right)\right]^{1/2}}, \tag{6.13}$$

where S_p^2, the pooled sample variance, is given by

$$S_p^2 = \frac{\sum_{i=1}^{2} \sum_{j=1}^{n_i} (x_{ij} - \bar{x}_i)^2}{n_1 + n_2 - 2}, \tag{6.14}$$

has a Student t distribution with $v = n_1 + n_2 - 2$ degrees of freedom.

Proof. Equation (6.14) may be written

$$S_p^2 = \frac{(n_1 - 1)s_1^2 + (n_2 - 1)s_2^2}{n_1 + n_2 - 2}.$$

Thus, by Theorem 6.2 and the additive property of χ^2, the quantity

$$w = S_p^2(n_1 + n_2 - 2)/\sigma^2, \tag{6.15}$$

is distributed as χ^2 with $(n_1 + n_2 - 2)$ degrees of freedom. Furthermore, we know, from Eqns (5.31) and (5.32), that $\bar{x} = \bar{x}_1 - \bar{x}_2$ is normally distributed with mean $\mu = u_1 - \mu_2$ and variance

$$\sigma_d^2 = \frac{\sigma^2}{n_1} + \frac{\sigma^2}{n_2}.$$

Thus the quantity

$$u = \frac{(\bar{x}_1 - \bar{x}_2) - (\mu_1 - \mu_2)}{\left[\sigma^2\left(\frac{1}{n_1} + \frac{1}{n_2}\right)\right]^{1/2}} = \frac{\bar{x} - \mu}{\sigma_d}, \tag{6.16}$$

is normally distribtued with mean zero and unit variance. Now, by Theorem 6.3, $\bar{x}_i$ and s_i^2 $(i = 1, 2)$ are independent random variables, and hence $\bar{x}$ and μ are independent random variables. Thus, by Theorem 6.4, the quantity

$$t = \frac{u}{[w/(n_1 + n_2 - 2)]^{1/2}}, \tag{6.17}$$

has a t distribution with $(n_1 + n_2 - 2)$ degrees of freedom. Substituting (6.15) and (6.16) into (6.17) gives (6.13) and completes the proof.

Like the χ^2 distribution the Student t distribution is also a one-parameter family of curves. Tables of percentage points are given in Appendix D, and in using them one can use the fact that

$$P[t < -t_\alpha(\nu)] = P[t > t_\alpha(\nu)] = \alpha,$$

since the distribution is symmetric about $t = 0$. The percentage points are also shown graphically in Fig. (6.3).

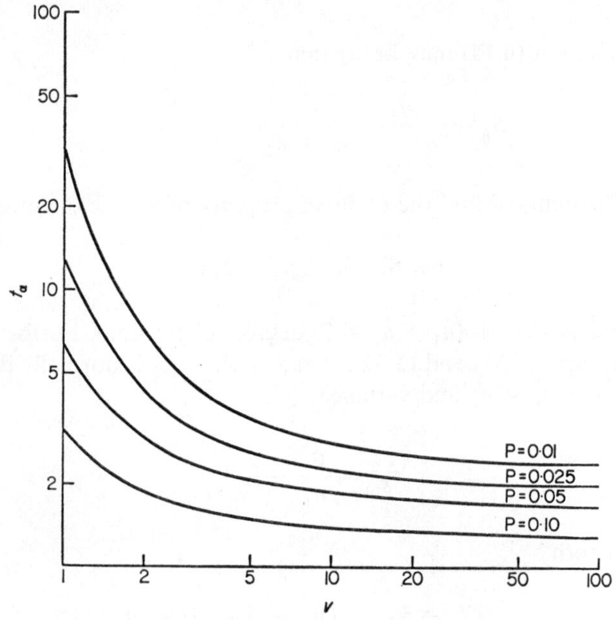

FIG. 6.3. Percentage points of the Student t distribution. $P = P[t \geqslant t_\alpha]$.

6.3 *F* distribution

The F distribution is designed for use in situations where we wish to compare two variances, or more than two means, situations for which the χ^2 and Student t distributions are not appropriate.

We will begin by constructing the form of the F distribution.

THEOREM 6.8. *Let the two independent random variables χ_i^2 $(i = 1, 2)$ be distributed as χ^2 with ν_i degrees of freedom. Then the statistic*

$$F \equiv F(\nu_1, \nu_2) = \frac{\chi_1^2/\nu_1}{\chi_2^2/\nu_2}, \tag{6.18}$$

is distributed with density function

$$f(F;\nu_1,\nu_2)=\frac{\Gamma\left(\dfrac{\nu_1+\nu_2}{2}\right)}{\Gamma(\nu_1/2)\Gamma(\nu_2/2)}\left(\frac{\nu_1}{\nu_2}\right)^{\nu_1/2}\frac{F^{(\nu_1-2)/2}}{\left(1+\dfrac{\nu_1}{\nu_2}F\right)^{(\nu_1+\nu_2)/2}},\quad F\geqslant 0,\quad (6.19)$$

mean

$$\mu=\frac{\nu_2}{\nu_2-2},\qquad \nu_2>2,$$

and variance

$$\sigma^2=\frac{2\nu_2{}^2(\nu_1+\nu_2-2)}{\nu_1(\nu_2-2)^2\,(\nu_2-4)},\qquad \nu_2>4.$$

Proof. Let

$$u=\chi_1{}^2;\qquad v=\chi_2{}^2,$$

then by Theorem 6.1 the joint density function of u and v is

$$g(u,v)=\frac{u^{(\nu_1-2)/2}\,v^{(\nu_2-2)/2}}{\Gamma(\nu_1/2)\Gamma(\nu_2/2)2^{(\nu_1+\nu_2)/2}}\exp\left[-\tfrac{1}{2}(u+v)\right].$$

Substituting

$$u=\left(\frac{\nu_1}{\nu_2}\right)vF,$$

gives the joint density of F and v as

$$f(F,v)=\frac{v^{(\nu_2-2)/2}}{\Gamma(\nu_1/2)\Gamma(\nu_2/2)2^{(\nu_1+\nu_2)/2}}\left(\frac{\nu_2 v}{\nu_1}\right)\times\left(\frac{\nu_1 vF}{\nu_2}\right)^{(\nu_1-2)/2}\exp\left[-\frac{v}{2}\left(1+\frac{\nu_1}{\nu_2}F\right)\right].$$

To obtain the density function of F we integrate out the dependence on v. Thus

$$f(F;\nu_1,\nu_2)=\frac{F^{(\nu_1-2)/2}}{\Gamma(\nu_1/2)\Gamma(\nu_2/2)2^{(\nu_1+\nu_2)/2}}\left(\frac{\nu_1}{\nu_2}\right)^{\nu_1/2}I(F;\nu_1,\nu_2),$$

where

$$I(F; \nu_1, \nu_2) = \int_0^\infty v^{(\nu_1+\nu_2-2)/2} \exp\left[-\frac{v}{2}\left(1 + \frac{\nu_1}{\nu_2}F\right)\right] dv$$

$$= \frac{\Gamma[(\nu_1 + \nu_2)/2]2^{(\nu_1+\nu_2)/2}}{(1 + \nu_1 F/\nu_2)^{(\nu_1+\nu_2)/2}}.$$

Equation (6.19) follows directly.

The m.g.f. may be deduced in the usual way. One finds that moments of order r exist only for $2r < \nu_2$, and are given by

$$\mu_r' = \left(\frac{\nu_2}{\nu_1}\right)^r \frac{\Gamma(r + \nu_1/2)\Gamma(\nu_2/2 - r)}{\Gamma(\nu_1/2)\Gamma(\nu_2/2)}. \tag{6.20}$$

The mean and variance follow directly. By using (6.20) to calculate β_1 it can be shown that the F distribution is always skewed.

The density function of the F distribution is more complicated than that of either the χ^2 or t distribution in being a two-parameter family of curves.

Percentage points are defined in the same way as for the χ^2 distribution. Thus

$$P[F \geqslant F_\alpha] = \alpha = \int_{F_\alpha}^\infty dF\, f(F; \nu_1, \nu_2).$$

Right-tail percentage points may be obtained from the tables in Appendix D, and should the left-tail percentage points be needed they may be obtained from the relation

$$F_{1-\alpha}(\nu_1, \nu_2) = [F_\alpha(\nu_2, \nu_1)]^{-1}.$$

The percentage points for $P = 0.05$ are also shown graphically in Fig. 6.4.

6.4 Relation between χ^2, t and F distributions

The F distribution is related in a simple way to the χ^2 and Student t distributions, as follows.

It is easy to show that as $\nu \to \infty$, $P[|\chi^2/\nu - 1|] \to 0$. Thus

$$F(\nu_1, \infty) = \frac{\chi_1^2}{\nu_1}. \tag{6.21}$$

Thus, the distribution of χ_1^2/ν_1 with ν_1 degrees of freedom is a special case of the F distribution with ν_1 and ∞ degrees of freedom. Thus for any α we have

$$F_\alpha(\nu_1, \infty) = \frac{\chi_\alpha^2(\nu_1)}{\nu_1}, \tag{6.22}$$

which may be directly verified by the use of a set of tables. If we consider the limit as $\nu_1 \to \infty$ then we have

$$F(\infty, \nu_2) = \frac{\nu_2}{\chi^2(\nu_2)}, \tag{6.23}$$

and

$$F_\alpha(\infty, \nu_2) = \frac{\nu_2}{\chi_{1-\alpha}^2(\nu_2)}. \tag{6.24}$$

Thus the left-tail percentage points of the χ^2/ν distribution are special cases of the right-tail percentage points of $F(\infty, \nu)$.

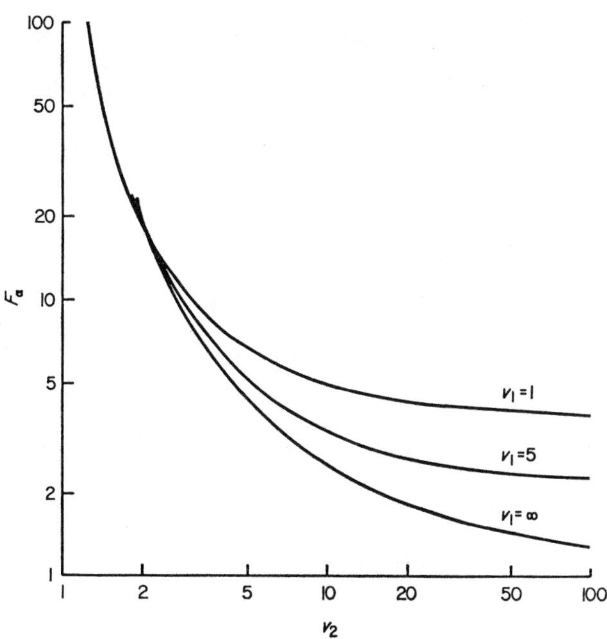

Fig. 6.4. Percentage points of the F distribution for $P = P[F \geqslant F_\alpha] = 0{\cdot}05$.

To relate the F-distribution to the Student t distribution we note that when $\nu_1 = 1$, $\chi^2/\nu_1 = u^2$, where u is a standard normal variate. Thus we may write

$$F(1, \nu_2) = \frac{u^2}{(\chi_2{}^2/\nu_2)}. \tag{6.25}$$

Now from Theorem 6.4

$$t = \frac{u}{(\chi_2{}^2/\nu_2)^{1/2}}, \tag{6.26}$$

is distributed as the Student t distribution with ν_2 degrees of freedom, and so we may write (6.25) as

$$F(1, \nu_2) = t^2(\nu_2). \tag{6.27}$$

Using (6.27) we may rewrite

$$P[F(1, \nu_2) < F_\alpha(1, \nu_2)] = 1 - \alpha$$

as

$$P[-(F_\alpha(1, \nu_2))^{1/2} < t(\nu_2) < (F_\alpha(1, \nu_2))^{1/2}] = 1 - \alpha,$$

and using the symmetry of the t distribution about $t = 0$ we have

$$P[t(\nu_2) < -(F_\alpha(1, \nu_2))^{1/2}] = P[t(\nu_2) > (F_\alpha(1, \nu_2))^{1/2}] = \alpha/2. \tag{6.28}$$

But

$$P[t(\nu_2) > t_{\alpha/2}(\nu_2)] = \alpha/2,$$

and so

$$t_{\alpha/2}(\nu_2) = [F_\alpha(1, \nu_2)]^{1/2},$$

or

$$F_\alpha(1, \nu_2) = t_{\alpha/2}{}^2(\nu_2). \tag{6.29}$$

Similarly, we can show that for $\nu_2 = 1$

$$F(\nu_1, 1) = [t^2(\nu_1)]^{-1}, \tag{6.30}$$

and

$$F_\alpha(\nu_1, 1) = [t_{(1+\alpha)/2}{}^2(\nu_1)]^{-1}. \tag{6.31}$$

Finally, if $\nu_1 = 1$ and $\nu_2 \to \infty$

$$F_\alpha(1, \infty) = u_{\alpha/2}^{\ 2}, \tag{6.32}$$

and if $\nu_2 = 1$ and $\nu_1 \to \infty$

$$F_\alpha(\infty, 1) = [u_{(1+\alpha)/2}^{\ 2}]^{-1}, \tag{6.33}$$

where u_α is a point of the standard normal variate such that

$$P[u > u_\alpha] = \alpha.$$

The above results are summarized in Table (6.2).

TABLE 6.2 Percentage points F_α of the $F(\nu_1, \nu_2)$ distribution and their relation to the χ^2 and Student t distributions

	ν_1		
ν_2	1	ν_1	∞
1	$t^2_{\alpha/2}(1) = \dfrac{1}{t_{(1+\alpha)/2}^{\ 2}(1)}$	$\dfrac{1}{t_{(1+\alpha)/2}^{\ 2}(\nu_1)}$	$\dfrac{1}{u_{(1+\alpha)/2}^{\ 2}}$
ν_2	$t_{\alpha/2}^{\ 2}(\nu_2)$	$F_\alpha(\nu_1, \nu_2)$	$\dfrac{\nu_2}{\chi_{1-\alpha}^{\ 2}(\nu_2)}$
∞	$u_{\alpha/2}^{\ 2}$	$\dfrac{\chi_\alpha^{\ 2}(\nu_1)}{\nu_1}$	1

7 Estimation I: Maximum Likelihood

In previous chapters we have encountered the problem of estimating from a sample the values of the parameters of a population. For example, we have used the sample mean as defined by Eqn (5.1) as an estimate of the population mean, and we have seen that this choice is supported by the Laws of Large Numbers. However, in the case of the variance, we did not define the sample variance by analogy with the population variance (*cf.* Eqns (5.2) and (3.3)). It is clearly time to consider the whole question of estimation more carefully, and in this section we will discuss firstly the properties required of methods of estimation in general, and then consider estimation by the so-called Maximum Likelihood Method. The discussion will be extended in Chapters 8 and 9 to include estimation by several other methods.

7.1 Properties of point estimators

Firstly, it is necessary to consider rather more closely what we mean by "estimation", and for this purpose it is useful to distinguish between the terms "estimator" and "estimate". By the former we mean the method of estimation, and by the latter we mean the value to which it gives rise in a particular case. The estimator is a random variable (e.g. the sample mean) and gives rise to a population of estimates, so the merit of an estimator is to be judged by the quality of this population, and not by the value of a particular estimate. To be a suitable estimator a quantity must satisfy certain criteria and these are given below.

It is intuitively obvious that a desirable property of an estimator is that, as the sample size increases, the estimate tends to the value of the population parameter. Any other result is clearly misleading. This property is known as *consistency* and is defined as follows.

DEFINITION 7.1. An estimator θ_n, computed from a sample of size n, is said to be a *consistent* estimator of a population parameter $\bar{\theta}$ if, for any positive ε and η, arbitrarily small, there exists some N such that

$$P[|\theta_n - \bar{\theta}| < \varepsilon] > 1 - \eta. \tag{7.1}$$

In these circumstances θ_n is said to *converge in probability* to $\bar{\theta}$. Thus θ_n is a consistent estimator of $\bar{\theta}$ if it converges in probability to $\bar{\theta}$.

EXAMPLE 7.1. Consider the Cauchy distribution, which we have defined in Section 4.4.

$$f(x;\theta) = \frac{1}{\pi} \cdot \frac{1}{1 + (x - \theta)^2}.$$

The sample mean

$$\bar{x} = \frac{1}{n} \sum_{i=1}^{n} x_i,$$

has a distribution which is also of the Cauchy form, independent of sample size (*cf.* Example 5.1). Thus the sample mean cannot converge in probability to θ (or any constant), and hence is not a consistent estimator of θ.

The property of consistency tells us the asymptotic ($n \to \infty$) behaviour of a suitable estimator. Having found such an estimator it is clear that we can generate an infinity of other estimators

$$\theta_n' = f(n)\theta_n, \tag{7.2}$$

provided

$$\lim_{n \to \infty} f(n) = 1. \tag{7.3}$$

However, we may further restrict the possible estimators by requiring that for *all n* the expected value of θ_n is $\bar{\theta}$. Such an estimator is called *unbiased.*

DEFINITION 7.2. An estimator θ_n, computed from a sample of size n, is said to be an *unbiased* estimator of a population parameter $\bar{\theta}$ if

$$E[\theta_n] = \bar{\theta}, \tag{7.4}$$

for all n.

EXAMPLE 7.2. By applying Eqn (7.4) to Eqn (5.1) we can trivially show that the sample mean $\bar{x}$ is an unbiased estimator of the population mean of Eqn (3.1). If we apply (7.4) to the definition of the sample variance of Eqn (5.2) we have

$$E\left[\frac{1}{n-1} \sum_{i=1}^{n} (x_i - \bar{x})^2\right] = \frac{1}{n-1} E\left[\sum_{i=1}^{n} \left(x_i - \frac{1}{n} \sum_{j=1}^{n} x_j\right)^2\right]$$

$$= \frac{1}{n-1} E\left[\frac{n-1}{n} \sum_{i=1}^{n} (x_i)^2 - \frac{1}{n} \sum_{i \neq j}^{n} x_i x_j\right]$$

$$= \mu_2' - (\mu_1')^2 = \sigma^2. \tag{7.5}$$

Thus the presence of the factor $1/(n-1)$ in the definition of the sample variance is to ensure that

$$E[s^2] = \sigma^2, \tag{7.6}$$

i.e. that s^2 is an unbiased estimator of σ^2. We will always use Eqn (5.2) as the definition of the sample variance. However some authors prefer to work with the biased estimator

$$s'^2 = \frac{1}{n} \sum_{i=1}^{n} (x_i - \bar{x})^2,$$

and so care should be taken when consulting different sources.

The requirements of consistency and lack of bias alone do not produce unique estimators. For example, one can easily show that the sample mean is a consistent and unbiased estimator of the mean of a normal population with known variance. But the same is true of the sample median. Thus we must impose further restrictions if uniqueness is required. One of these is known as the *efficiency* of an estimator.

An unbiased estimator with a smaller variance will produce estimates more closely grouped round the population value $\bar{\theta}$. An estimator with a smaller variance is said to be more *efficient* than one with a larger variance.

DEFINITION 7.3. If two consistent estimators θ_1 and θ_2, both calculated from a sample of size n, have $\operatorname{var}\theta_1 < \operatorname{var}\theta_2$ then θ_1 is said to be more *efficient* than θ_2 for samples of size n.

EXAMPLE 7.3. For the normal distribution we have, from Theorem 5.2, for any n

$$\operatorname{var}(\text{mean}) = \sigma^2/n.$$

But for large n

$$\operatorname{var}(\text{median}) = \frac{\pi\sigma^2}{2n} > \frac{\sigma^2}{n}.$$

Thus the mean is the more efficient estimator, at least for large n. (In fact this result is true for all n). Consistent estimators whose sampling variance for large samples are less than that of any other such estimator are called *most efficient*. Such estimators serve to define a scale of efficiency. Thus if θ_2 has variance v_2 and θ_1, the most-efficient estimator, has variance v_1, then the efficiency of θ_2 is defined as

$$E_2 = \frac{v_1}{v_2}. \tag{7.7}$$

It may still be that there exist several consistent estimators θ for a population parameter $\bar{\theta}$. Can one choose a "best" estimator from amongst them? The criterion of efficiency alone is not enough, since it is possible that one estimator θ_n, which is biased, is consistently closer to $\bar{\theta}$ than an unbiased estimator θ_n'. In this case the quantity to consider is not the variance but the second moment of θ_n about $\bar{\theta}$, which is

$$E[(\theta_n - \bar{\theta})^2].$$

When

$$\bar{\theta} = E[\theta_n],$$

then

$$E[(\theta_n - \bar{\theta})^2] = \text{var}(\theta_n).$$

Thus we shall define θ_n to be a *best estimator* of the parameter $\bar{\theta}$ if

$$E[(\theta_n - \bar{\theta})^2] \leqslant E[(\theta_n' - \bar{\theta})^2],$$

where θ_n' is any other estimator of $\bar{\theta}$.

The properties of estimators considered up to now give a good idea of the desirable properties of estimators. However, there is a more general criterion which we will now consider. Consider the case of estimating a parameter $\bar{\theta}$. Let

$$f(\theta_1, \theta_2, \ldots, \theta_r; \bar{\theta}),$$

be the joint density function of r independent estimators θ_i $(i = 1, 2, \ldots, r)$. Then, from the definition of the multivariate conditional density Eqn (3.30), we have

$$f(\theta_1, \theta_2, \ldots, \theta_r; \bar{\theta}) = f^M(\theta_1; \bar{\theta}) f^C(\theta_2, \theta_3, \ldots, \theta_r; \bar{\theta}|\theta_1), \tag{7.8}$$

where $f^M(\theta_1; \bar{\theta})$ is the marginal density of θ_1 and $f^C(\theta_2, \ldots, \theta_r; \bar{\theta}|\theta_1)$ is the conditional density of all the other θ_i, given θ_1. Now if f^C is independent of $\bar{\theta}$ then clearly once θ_1 is specified the other estimators contribute nothing further to the problem of estimating $\bar{\theta}$, i.e. θ_1 contains *all* the information about $\bar{\theta}$. In these circumstances θ_1 is called a *sufficient* statistic for $\bar{\theta}$. It is more convenient in practice to write (7.8) as a condition on the likelihood function.

Definition 7.4. Let $f(x; \bar{\theta})$ denote the density function of a random variable x, where the form of f is known but not the value of $\bar{\theta}$, which is to be estimated.

Let $x_1, x_2, ..., x_n$ be a random sample of size n. The joint density function $f(x_1, x_2, ..., x_n; \bar{\theta})$ of the independent random variables $x_1, x_2, ..., x_n$ is given by

$$f(x_1, x_2, ..., x_n; \bar{\theta}) = \prod_{i=1}^{n} f(x_i; \bar{\theta}), \tag{7.9}$$

where $f(x_i; \bar{\theta})$ is the density function for the ith random variable. $f(x_1, x_2, ..., x_n)$ is called the *likelihood function* of $\bar{\theta}$ and is written

$$L(x_1, x_2, ..., x_n; \bar{\theta}) = \prod_{i=1}^{n} f(x_i; \bar{\theta}). \tag{7.10}$$

If L is expressible in the form

$$L(x_1, x_2, ..., x_n; \bar{\theta}) = L_1(\theta, \bar{\theta}) L_2(x_1, x_2, ..., x_n), \tag{7.11}$$

where L_1 does not contain the x's other than in the form θ, and L_2 is independent of $\bar{\theta}$, then θ is a *sufficient* statistic for the estimation of $\bar{\theta}$.

EXAMPLE 7.4. We will find a sufficient statistic for estimating the value of the variance of a normal distribution with zero mean, i.e.

$$f(x) = \frac{1}{\sqrt{(2\pi)}} \frac{1}{\sigma} \exp\left[\frac{-x^2}{2\sigma^2}\right].$$

From Eqn (7.9), the likelihood function is

$$L(x_1, x_2, ..., x_n; \sigma^2) = \left(\frac{1}{\sigma\sqrt{2\pi}}\right)^n \exp\left(-\frac{1}{2\sigma^2}\sum_{i=1}^{n} x_i^2\right).$$

If we let $L_2 = 1$ in Eqn (7.11) then we have $L_1 = L$, and L_1 is a function of the sample x_i only in terms of Σx_i^2. Thus, by Definition 7.4, Σx_i^2 is a sufficient estimator for σ^2.

7.2 Maximum likelihood method

Of all the possible methods of parameter estimation that of maximum likelihood is, in a sense to be discussed below, the most general, and is widely used in practice. We have briefly mentioned it in Chapter 2, but in the present chapter we will consider the method in more detail.

7.2.1. ESTIMATION OF A SINGLE PARAMETER

The likelihood function has been defined in Eqn (7.10). If we suppress the dependence of L on x_i, then for a sample of size n

$$L(\theta) = \prod_{i=1}^{n} f(x_i; \bar{\theta}), \tag{7.12}$$

where $f(x; \bar{\theta})$ is the frequency function of the parent population. The maximum likelihood estimator is defined as follows.

DEFINITION 7.5. The maximum likelihood estimator (M.L.E.) of a population parameter $\bar{\theta}$ is that statistic $\hat{\theta}$ (we shall always use the symbol $\wedge$ to denote an estimate of a parameter) which maximizes $L(\theta)$ for variations of θ, i.e. the solution (if it exists) of the equations

$$\frac{\partial L(\theta)}{\partial \theta} = 0; \qquad \frac{\partial^2 L(\theta)}{\partial \theta^2} < 0. \tag{7.13}$$

Since $L(\theta) > 0$ the first equation is equivalent to

$$\frac{1}{L}\frac{\partial L(\theta)}{\partial \theta} = \frac{\partial}{\partial \theta}\ln L(\theta) = 0, \tag{7.14}$$

which is the form more often used in practice. It is clear from (7.13) that the solution obtained by estimating the parameter θ is the same as that obtained by estimating a function of θ, e.g. $F(\theta)$ since

$$\frac{\partial \ln L}{\partial \theta} = \frac{\partial \ln L(F)}{\partial F} \cdot \frac{\partial F}{\partial \theta}, \tag{7.15}$$

and the two sides of the equation vanish together.

The importance of M.L.E. stems from the following four theorems, which we state without proof.

THEOREM 7.1. *Maximum likelihood estimators are consistent.*

THEOREM 7.2. *Maximum likelihood estimators have a distribution which tends to normality for large samples.*

THEOREM 7.3. *Maximum likelihood estimators have minimum variance in the limit of large samples.*

THEOREM 7.4. *If a sufficient estimator for a parameter exists then it is a function of the maximum likelihood estimator.*

There are situations in which the above theorems do not hold and the M.L.E. gives a poor estimate for a parameter, but for the common distributions met in practice they are valid. Proofs of these theorems, together with exact statements on the range of their validity, may be found in the books of Kendall and Stuart, and Cramèr, cited in the Bibliography.

The likelihood function $L(\theta)$ may be *formally* regarded as a probability density function for the parameter θ viewed as a random variable. Thus we may define the variance of the estimator as

$$\operatorname{var}\hat{\theta} = \int_{-\infty}^{\infty} (\theta - \hat{\theta})^2 L(\theta) d\theta \Big/ \int_{-\infty}^{\infty} L(\theta) d\theta, \tag{7.16}$$

and, by analogy with the work of Section 5.3, an estimate from experimental data would be quoted as

$$\theta = \hat{\theta}_e \pm \Delta\hat{\theta},$$

where

$$\Delta\hat{\theta} = (\operatorname{var}\hat{\theta}_e)^{\frac{1}{2}}.$$

From Theorem 7.2, it follows that, for large samples, the form of $L(\theta)$ is

$$L(\theta) = \frac{1}{(2\pi v)^{\frac{1}{2}}} \exp\left[-\tfrac{1}{2}\cdot\frac{(\theta - \hat{\theta})^2}{v}\right], \tag{7.17}$$

where

$$v = \operatorname{var}\theta.$$

Now

$$\ln L(\theta) = -\ln\left[(2\pi v)^{\frac{1}{2}}\right] - \tfrac{1}{2}\frac{(\theta - \hat{\theta})^2}{v}, \tag{7.18}$$

and

$$\frac{\partial^2 \ln L(\theta)}{\partial \theta^2} = \frac{-1}{v}.$$

Thus

$$\operatorname{var}\hat{\theta} = \left[-\frac{\partial^2 \ln L(\theta)}{\partial \theta^2}\right]^{-1}\Bigg|_{\theta = \hat{\theta}}. \tag{7.19}$$

This is another commonly used form for the variance of an estimate. We shall illustrate the use of the maximum likelihood method for one parameter by two examples.

EXAMPLE 7.5. We shall estimate the parameter μ in the normal population given by

$$f(x;\mu,\sigma^2) = \frac{1}{(2\pi\sigma^2)^{\frac{1}{2}}} \exp\left[-\tfrac{1}{2}\left(\frac{x-\mu}{\sigma}\right)^2\right], \tag{7.20}$$

where σ is known and $-\infty \leqslant x \leqslant \infty$, and also calculate its variance.

From (7.12) we have

$$\ln L(\mu) = -n\ln[(2\pi\sigma^2)^{\frac{1}{2}}] - \frac{1}{2\sigma^2}\sum_{i=1}^{n}(x_i-\mu)^2,$$

and hence the M.L.E. of μ is the solution of

$$\frac{\partial \ln L(\mu)}{\partial \mu} = \frac{1}{\sigma^2}\sum_{i=1}^{n}(x_i-\mu) = 0,$$

i.e.

$$\hat{\mu} = \frac{1}{n}\sum_{i=1}^{n} x_i = \bar{x}. \tag{7.21}$$

Thus the sample mean $\bar{x}$ is the M.L.E. of the parameter μ.

From (7.19) the variance of $\hat{\mu}$ is given by

$$\begin{aligned}\operatorname{var}\hat{\mu} &= \left[-\frac{1}{2\sigma^2}\sum_{i=1}^{n}\frac{\partial^2}{\partial\mu^2}(x_i-\mu)^2\right]^{-1}\Bigg|_{\mu=\hat{\mu}} \\ &= \sigma^2/n,\end{aligned}$$

as expected from the result of Theorem 5.2.

EXAMPLE 7.6. As a more practical example we shall consider estimation of the parameter μ in the same normal population, but now for a set of experimental observations of the same quantity x_i made with associated errors Δx_i. The density function is

$$f(x,\Delta x;\mu) = \frac{1}{\sqrt{(2\pi)}\Delta x}\exp\left[-\tfrac{1}{2}\left(\frac{x-\mu}{\Delta x}\right)^2\right], \tag{7.22}$$

from which

$$\ln L(\mu) = -\ln\left[(2\pi)^{\frac{1}{2}}\sum_{i=1}^{n}\Delta x_i\right] - \tfrac{1}{2}\sum_{i=1}^{n}\left(\frac{x_i-\mu}{\Delta x_i}\right)^2,$$

and

$$\frac{\partial \ln L(\mu)}{\partial \mu} = \sum_{i=1}^{n} \left[\frac{x_i - \mu}{(\Delta x_i)^2}\right]. \tag{7.23}$$

Setting (7.23) to zero gives

$$\hat{\mu} = \sum_{i=1}^{n} (x_i/\Delta x_i^2) \Big/ \sum_{i=1}^{n} (1/\Delta x_i^2), \tag{7.24}$$

a result which is often called the *weighted mean* of a set of observations.

The variance of $\hat{\mu}$ may be found from (7.23) and (7.19) directly. It is

$$\text{var } \hat{\mu} = (\Delta\hat{\mu})^2 = \left[\sum_{i=1}^{n} \left(\frac{1}{\Delta x_i}\right)^2\right]^{-1}. \tag{7.25}$$

If an experiment has "good statistics" then the likelihood function will indeed be a close approximation to a normal distribution. However, many effects may be present which produce a function which is clearly not normal. In this case the use of Eqn (7.19) usually produces an underestimate for $\Delta\hat{\theta}$. A more realistic estimate of $\Delta\hat{\theta}$ is to average $\partial^2 \ln L(\theta)/\partial\theta^2$ over the likelihood function, i.e. take

$$\Delta\hat{\theta} = \left\{ \frac{\displaystyle\int_{-\infty}^{\infty} \left(-\frac{\partial^2 \ln L(\theta)}{\partial \theta^2}\right) L(\theta) d\theta}{\displaystyle\int_{-\infty}^{\infty} L(\theta) d\theta} \right\}^{-1/2}, \tag{7.26}$$

or alternatively, one could plot $L(\theta)$, and then find the two values of θ where $L(\theta)$ had fallen by a factor $e^{\frac{1}{2}}$ of its maximum, i.e. the two values which would correspond to $\mu \pm \sigma$ for a normal distribution of mean μ and variance σ^2.

Another useful formula for $\Delta\hat{\theta}$ may be derived for situations where one wants to answer the question; how many data are required to establish a particular result to within a specified accuracy? The problem is to find a value for $\partial^2 \ln L(\theta)/\partial\theta^2$ averaged over many repeated experiments consisting of n events each. Since

$$\ln L(x; \theta) = \sum_{i=1}^{n} \ln f(x_i; \theta),$$

we have

$$\overline{\left(\frac{\partial^2 \ln L(\theta)}{\partial\theta^2}\right)} = n\int \frac{\partial^2 \ln f(x;\theta)}{\partial\theta^2} f(x;\theta)dx$$

$$= nE\left[\frac{\partial^2 \ln f(x;\theta)}{\partial\theta^2}\right]. \tag{7.27}$$

This form may be used in (7.19) directly, or it may be expressed in terms of first derivatives by

$$\int \frac{\partial^2 \ln f(x;\theta)}{\partial\theta^2} f(x;\theta)d\theta = -\int \left(\frac{\partial f(x;\theta)}{\partial\theta}\right)^2 \frac{dx}{f(x;\theta)}. \tag{7.28}$$

Using (7.27) and (7.28) in (7.19) gives

$$\Delta\hat{\theta} = \frac{1}{n^{\frac{1}{2}}}\left[\int \left(\frac{\partial f(x;\theta)}{\partial\theta}\right)^2 \frac{1}{f(x;\theta)} dx\right]^{-\frac{1}{2}}. \tag{7.29}$$

EXAMPLE 7.7. Consider a distribution with density function

$$f(x;\theta) = \tfrac{1}{2}(1+\theta x), \qquad -1 \leqslant x \leqslant 1. \tag{7.30}$$

We have

$$\frac{\partial f(x;\theta)}{\partial\theta} = \frac{x}{2},$$

and

$$\int_{-1}^{1} \left(\frac{\partial f(x;\theta)}{\partial\theta}\right)^2 \frac{1}{f(x;\theta)} dx = \frac{1}{2\theta^3}\left[\ln\left(\frac{1+\theta}{1-\theta}\right) - 2\theta\right]. \tag{7.31}$$

Thus from (7.29)

$$\left(\frac{\Delta\theta}{\theta}\right) = \frac{1}{n^{\frac{1}{2}}}\left[\frac{2\theta}{\ln\left(\dfrac{1+\theta}{1-\theta}\right) - 2\theta}\right]^{1/2}. \tag{7.32}$$

Suppose we wish to establish how many events would be required to obtain $(\Delta\theta/\theta) = 0{\cdot}01$ for $\theta = 0{\cdot}5$. Substituting these numbers into (7.32) gives directly $n \simeq 1{\cdot}1 \times 10^5$ events. In particular, Eqn (7.29) shows that to increase the precision of the experiment n-fold requires n^2 times as many events.

All the results of this subsection may be generalized in a straightforward way to the case of estimating a single parameter from a multivariate distribution.

7.2.2. Simultaneous Estimation of Several Parameters

If we wish to estimate simultaneously several parameters then the preceding results generalize in a straightforward way. Thus, the maximum likelihood equation now becomes the set of simultaneous equations

$$\frac{\partial \ln L(\theta_1, \theta_2, \ldots, \theta_i, \ldots, \theta_n)}{\partial \theta_i} = 0, \qquad i = 1, 2, \ldots, n. \tag{7.33}$$

Also the results of Theorems (7.1)–(7.4) hold. We shall state explicitly the generalization of Theorem 7.2.

Theorem 7.5. *The M.L.E.* $\hat{\theta}_i \, (i = 1, 2, \ldots, p)$ *for the parameters of a density* $f(x; \theta_1, \ldots, \theta_p)$ *from samples of size n are, for large samples, approximately distributed as the multivariate normal distribution with means* $\theta_1, \ldots, \theta_p$ *and with a variance matrix* $\mathbf{V}$ *where*

$$M_{ij} = (V^{-1})_{ij} = -nE\left[\frac{\partial^2 \ln f(x; \theta_1, \ldots, \theta_p)}{\partial \theta_i \partial \theta_j}\right]. \tag{7.34}$$

To illustrate the use of (7.33) and (7.34) we shall give the following example.

Example 7.8. Consider the normal population

$$f(x; \mu, \sigma) = \frac{1}{(2\pi\sigma^2)^{\frac{1}{2}}} \exp\left[-\tfrac{1}{2}\left(\frac{x - \mu}{\sigma}\right)^2\right], \tag{7.35}$$

where both μ and σ are to be estimated. From (7.33) we have

$$\frac{\partial \ln L(\mu, \sigma)}{\partial \mu} = \frac{1}{\sigma^2} \sum_{j=1}^{n} (x_j - \mu) = 0,$$

$$\frac{\partial \ln L(\mu, \sigma)}{\partial \sigma} = \frac{1}{2\sigma^4} \sum_{j=1}^{n} (x_j - \mu)^2 - \frac{n}{2\sigma^2} = 0,$$

giving

$$\hat{\mu} = \bar{x}; \quad \hat{\sigma}^2 = \frac{1}{n} \sum_{j=1}^{n} (x_j - \bar{x})^2. \tag{7.36}$$

Note that $\hat{\sigma}^2$ is a biased estimator of σ^2 [*cf.* Eqn (7.2)]. This is often the case with M.L.E.'s, but fortunately there usually exists a constant c, in this case $n/(n-1)$, such that $c\hat{\sigma}^2$ is unbiased.

From Theorem (7.5) the two estimators μ and $\hat{\sigma}$ are approximately normally distributed with means μ and σ, and with a matrix **M** given by Eqn (7.34). Using (7.34) we have, with $\mu = \theta_1$ and $\sigma = \theta_2$,

$$M_{11} = -nE\left[-\frac{1}{\sigma^2}\right] = \frac{n}{\sigma^2},$$

$$M_{12} = M_{21} = -nE\left[-\frac{2(x-\mu)}{\sigma^3}\right] = 0,$$

$$M_{22} = -nE\left[-\frac{3(x-\mu)^2}{\sigma^4} + \frac{1}{\sigma^2}\right] = \frac{2n}{\sigma^2}.$$

Thus

$$V_{ij} = (M^{-1})_{ij} = \begin{pmatrix} \sigma^2/n & 0 \\ 0 & \sigma^2/2n \end{pmatrix}, \tag{7.37}$$

and the variance and covariances are given by

$$\frac{1}{n}\sigma_{ij} = V_{ij}. \tag{7.38}$$

Finally, from (4.16) the form of the distribution of the estimators is

$$S(\hat{\mu}, \hat{\sigma}) = \frac{\sqrt{2}n}{2\pi\sigma^2} \exp\left\{-\frac{n}{2}\left[2\left(\frac{\hat{\sigma}-\sigma}{\sigma}\right)^2 + \left(\frac{\hat{\mu}-\mu}{\sigma}\right)^2\right]\right\}. \tag{7.39}$$

There is one point which should be remarked about the simultaneous estimation of several parameters, which we shall illustrate by reference to the above example. If we know μ then estimation of σ^2 alone gives

$$\hat{\sigma}^2 = \frac{1}{n}\sum_{i=1}^{n}(x_i - \mu)^2, \tag{7.40}$$

which is not the same as Eqn (7.36) obtained from the simultaneous estimation of μ and σ^2. This is not surprising. However, from (7.36) we see that we can estimate μ, independent of any possible knowledge of σ^2, to be $\bar{x}$. Thus, if we now find the estimator of σ^2 maximizing the likelihood for all samples giving the estimated value of $\mu = \bar{x}$ it might be thought that Eqn (7.36) will result, whereas in fact in this latter case

$$\hat{\sigma}^2 = \frac{1}{n-1}\sum_{i=1}^{n}(x_i - \bar{x})^2. \tag{7.41}$$

The difference between (7.36) and (7.41) is that in the former case we have considered the variations of $\ln L(\mu, \sigma)$ over all samples of size n, whereas in the latter the constraint, $\Sigma(x) =$ constant, has been applied, and thus lowered the number of degrees of freedom by one. For large n, of course, the difference is of no importance, but nevertheless it is a useful reminder that every parameter estimated from the sample (i.e. every constraint applied) lowers the number of degrees of freedom by one.

The maximum likelihood method has the disadvantage that in order to estimate a parameter the form of the distribution must be known. Furthermore, it often happens that $L(\boldsymbol{\theta})$ is a highly non-linear function of the parameters $\boldsymbol{\theta}$, and so to maximize the likelihood may be a difficult problem. (In Appendix C we shall consider some methods which are useful for the numerical optimization of a function of several variables.) Finally, if the data under study are normally distributed then maximizing $\ln L(\theta)$ is equivalent to minimizing

$$\chi^2 = \sum_{i=1}^{n} \left(\frac{x_i - \mu_i}{\sigma_i} \right)^2,$$

which may be more useful in practice as we shall illustrate when we discuss the method of least-squares in Chapter 8.

We will conclude this section with a few brief remarks on the interpretation of maximum likelihood estimators. Bayes' Theorem tells us that maximizing the likelihood does not necessarily maximize the *a posteriori* probability of an event. This is only the case if the *a priori* probabilities are equal or somehow "smooth". Thus, maximum likelihood estimators (and, of course, other estimators) should always be interpreted in the light of prior knowledge. In Chapter 9 we shall show how such knowledge can formally be included in the estimation procedure. However, because it is difficult, in general, to reduce prior knowledge to the required form, the actual method of estimation is of little practical use. An alternative method is to form the product of the respective likelihood functions. This procedure is equivalent to forming a likelihood function for an experiment which includes all previous experiments.

8 Estimation II: Least-Squares Method

The method of least-squares is an application of minimum variance estimators (which we will meet again in Chapter 9) to the multivariate problem, and is widely used in situations where a functional form is known (or assumed) to exist between the observed quantities and the parameters to be estimated. The functional form may be dictated by the requirements of a theoretical model of the data, or may be chosen arbitrarily to provide a convenient interpolation formula for use in other calculations. We will firstly consider the technique for the situation where it is most useful, that where the data depend *linearly* on the parameters to be estimated. In this form the least-squares method is frequently used in curve-fitting problems.

8.1 Linear least-squares

Initially we shall formulate the method as a procedure for finding estimators $\hat{\theta}_i \, (i = 1, \ldots, p)$ of parameters $\theta_i \, (i = 1, \ldots, p)$ which minimize the function

$$S = \sum_{i=1}^{n} (y_i - \hat{\eta}_i)^2 = \sum_{i=1}^{n} r_i^2, \tag{8.1}$$

where

$$\hat{\eta}_i = f(x_{1i}, x_{2i}, \ldots, x_{ki}; \hat{\theta}_1, \ldots, \hat{\theta}_p), \qquad i = 1, 2, \ldots, n. \tag{8.2}$$

and $y_i, x_{1i}, x_{2i}, \ldots, x_{ki}$ denote the ith set of observations on $(k + 1)$ variables, of which only y_i is random. The relation

$$\hat{\eta} = f(x_1, x_2, \ldots, x_k; \hat{\theta}_1, \hat{\theta}_2, \ldots, \hat{\theta}_p), \tag{8.3}$$

is called the *equation of the regression curve of best fit*.

We shall consider the general case where the observations are correlated and have different "weights". Suppose we make observations of a quantity y which is a function $f(x; \theta_1, \ldots, \theta_p)$ of one variable x and p parameters $\theta_i \, (i = 1, 2, \ldots, p)$. (Note that x is *not* a random variable and f is *not* a density function). The observations y_i are made at points x_i and are subject to experimental errors e_i. If the n observations y_i depend *linearly* on the p parameters then the observational equations may be written

$$y_i = \sum_{k=1}^{p} \theta_k \phi_k(x_i) + e_i, \qquad i = 1, 2, \ldots, n \tag{8.4}$$

where $\phi_k(x)$ are any linearly independent functions of x. In matrix notation Eqn (8.4) may be written

$$\mathbf{Y} = \mathbf{\Phi\Theta} + \mathbf{E}, \tag{8.5}$$

where $\mathbf{Y}$ and $\mathbf{E}$ are $(n \times 1)$ column vectors, $\mathbf{\Theta}$ is a $(p \times 1)$ column vector, and $\mathbf{\Phi}$ is the $(n \times p)$ matrix

$$\mathbf{\Phi} = \begin{pmatrix} \phi_1(x_1) & \phi_2(x_1) & \cdots & \cdots & \phi_p(x_1) \\ \phi_1(x_2) & \phi_2(x_2) & \cdots & \cdots & \phi_p(x_2) \\ \cdots & \cdots & & & \cdots \\ \cdots & \cdots & & & \cdots \\ \phi_1(x_n) & \phi_2(x_n) & \cdots & \cdots & \phi_p(x_n) \end{pmatrix}.$$

The matrix $\mathbf{\Phi}$ is often known as the *design matrix.*

8.1.1. Solution for the Parameters

The problem is to obtain estimates $\hat{\theta}_k$ for the parameters. For $n = p$ a unique solution is obtainable directly from Eqn (8.5) by a simple matrix inversion, but for the more practical case $n > p$ the system of equations is overdetermined. In this situation no general unique solution exists, and so what we seek is a "best average solution" in some sense. Thus we seek to approximate the experimental points y_i by a series of degree p, i.e.

$$f_i \equiv f(x_i; \theta_1, \ldots, \theta_p) = \sum_{k=1}^{p} \theta_k \phi_k(x_i). \tag{8.6}$$

Since the experimental errors are assumed to be random we would expect them to have a joint distribution with zero mean, i.e.

$$E[\mathbf{Y}] \equiv \mathbf{Y}^0 = \mathbf{\Phi\Theta}, \tag{8.7}$$

and an associated variance matrix

$$V_{ij} = \begin{pmatrix} \sigma_1^2 & \sigma_{12} & \cdots & \cdots & \sigma_{1n} \\ \sigma_{12} & \sigma_2^2 & \cdots & \cdots & \sigma_{2n} \\ \vdots & \vdots & & & \vdots \\ \vdots & \vdots & & & \vdots \\ \sigma_{1n} & \sigma_{2n} & \cdots & \cdots & \sigma_n^2 \end{pmatrix}, \tag{8.8}$$

where

$$\sigma_i^2 = E[e_i^2] = \text{var}\,(y_i),$$

and

$$\sigma_{ij} = E[e_i e_j] = \text{cov}\,(y_i, y_j).$$

Note that we have only assumed that the population distribution of the errors has a *finite second moment*. In particular, it is *not* necessary to assume that the distribution is normal. However, *if* the errors are normally distributed, as is often the case in practice, then the method of least-squares gives the same results as the method of maximum likelihood.

The quantities r_i of Eqn (8.1) (called the *residuals*) are now replaced by

$$r_i \equiv y_i - \hat{f}_i = y_i - \sum_{k=1}^{p} \theta_k \phi_k(x_i), \tag{8.9}$$

and we will minimize

$$\begin{aligned} S &= \sum_{i=1}^{n} \sum_{j=1}^{n} r_i r_j V_{ij}^{-1}, \\ &= \mathbf{R}^T \mathbf{V}^{-1} \mathbf{R}, \end{aligned} \tag{8.10}$$

where $\mathbf{R}$ is an $(n \times 1)$ column vector of residuals.

It is convenient at this stage to assume that the variance matrix can be expressed in the form

$$\mathbf{V} = \sigma^2 \mathbf{W}^{-1}, \tag{8.11}$$

where σ^2 is a scale factor and $\mathbf{W}$ is the so-called *weight matrix* of the observations. In that case Eqn (8.10) becomes

$$S = (\mathbf{Y} - \mathbf{\Phi\Theta})^T \mathbf{W} (\mathbf{Y} - \mathbf{\Phi\Theta}) \frac{1}{\sigma^2}. \tag{8.12}$$

To minimize S with respect to $\mathbf{\Theta}$ we have

$$\frac{\partial S}{\partial \mathbf{\Theta}} = 0,$$

giving a solution

$$\hat{\mathbf{\Theta}} = (\mathbf{\Phi}^T\mathbf{W}\mathbf{\Phi})^{-1}\mathbf{\Phi}^T\mathbf{W}\mathbf{Y}, \tag{8.13}$$

or, in non-matrix notation,

$$\hat{\theta}_k = \sum_{l=1}^{p} (E^{-1})_{kl} \sum_{i=1}^{n}\sum_{j=1}^{n} \phi_l(x_i)W_{ij}y_j, \tag{8.14}$$

where

$$E_{kl} = \sum_{i=1}^{n}\sum_{j=1}^{n} \phi_k(x_i)W_{ij}\phi_l(x_j). \tag{8.15}$$

These are the so-called *normal equations* for the parameters. One point worth remarking about the normal equations is that they do not require a knowledge of σ^2, only the relative weight matrix $\mathbf{W}$ is required to estimate the parameters.

The estimates $\hat{\theta}_k$ of Eqn (8.14) have been obtained by minimizing the sum of the residuals, and although this has an intuitive geometrical appeal it still might be thought to be a rather arbitrary procedure. However, the importance of least-squares estimates stems from their "minimum-variance" properties which are summarized by the following theorem.

THEOREM 8.1. *The least-squares estimate $\hat{\theta}_k$ of the parameters θ_k is that estimate which minimizes the variance of any linear combination of the parameters.*

Proof. Consider the general linear sum of parameters

$$L = \mathbf{C}^T\mathbf{\Theta}, \tag{8.16}$$

where $\mathbf{C}$ is a $(p \times 1)$ vector of known constant coefficients. Let $\mathbf{G}$ be any $(n \times 1)$ vector such that

$$\mathbf{C}^T = \mathbf{G}^T\mathbf{\Phi}. \tag{8.17}$$

The problem of minimizing the variance of L is now equivalent to minimizing the variance of $\mathbf{G}^T\mathbf{Y}$ subject to the constraint (8.17). Now since $\mathbf{G}$ is a constant vector

$$\text{var}\,(\mathbf{G}^T\mathbf{Y}) = \mathbf{G}^T(\text{var}\,\mathbf{Y})\mathbf{G} = \mathbf{G}^T\mathbf{V}\mathbf{G}.$$

Thus we can construct a variational function

$$F = \mathbf{G}^T\mathbf{V}\mathbf{G} - \mathbf{\Lambda}^T(\mathbf{\Phi}^T\mathbf{G} - \mathbf{C}), \tag{8.18}$$

where $\mathbf{\Lambda}$ is a $(p \times 1)$ vector of Lagrange multipliers. Setting $\delta F = 0$ gives

$$\mathbf{G}^T = \mathbf{\Lambda}^T \mathbf{\Phi}^T \mathbf{V}^{-1}, \tag{8.19}$$

and so

$$\mathbf{\Lambda}^T = \mathbf{G}^T \mathbf{\Phi}(\mathbf{\Phi}^T \mathbf{V}^{-1} \mathbf{\Phi})^{-1}. \tag{8.20}$$

Eliminating $\mathbf{\Lambda}^T$ between (8.19) and (8.20) gives

$$\mathbf{G}^T = \mathbf{G}^T \mathbf{\Phi}(\mathbf{\Phi}^T \mathbf{V}^{-1} \mathbf{\Phi})^{-1} \mathbf{\Phi}^T \mathbf{V}^{-1}. \tag{8.21}$$

If we now multiply (8.21) on the right by $\mathbf{Y}$ and use Eqn (8.13) we have

$$\mathbf{G}^T \mathbf{Y} = (\mathbf{G}^T \mathbf{\Phi})\mathbf{\Theta} = \mathbf{C}^T \hat{\mathbf{\Theta}}. \tag{8.22}$$

Thus we have shown that the value of $\hat{\mathbf{\Theta}}$ which minimizes the variance of any linear combination of the parameters is the least-squares estimate $\hat{\mathbf{\Theta}}$. This result is originally due to Gauss.

8.1.2. Errors on the Parameter Estimates

Having obtained the least-squares estimates $\hat{\theta}_k$ we have now to consider their variances and covariances. Returning to the solution of the normal equations, we have

$$\hat{\mathbf{\Theta}} = (\mathbf{\Phi}^T \mathbf{W} \mathbf{\Phi})^{-1} \mathbf{\Phi}^T \mathbf{W} \mathbf{Y}. \tag{8.23}$$

Now, we have previously used the result, that for any linear combination of y_i, say $\mathbf{P}^T\mathbf{Y}$, with $\mathbf{P}$ a constant vector

$$\text{var}\,(\mathbf{P}^T \mathbf{Y}) = \mathbf{P}^T \text{var}\,(\mathbf{Y})\mathbf{P}, \tag{8.24}$$

which can easily be proved from the definition of the variance matrix. Thus, applying Eqn (8.24) to $\hat{\mathbf{\Theta}}$ as given by Eqn (8.23) we have

$$\text{var}\,(\hat{\mathbf{\Theta}}) = (\mathbf{\Phi}^T \mathbf{W} \mathbf{\Phi})^{-1} \mathbf{\Phi}^T \mathbf{W}\, \text{var}\,(\mathbf{Y}) \mathbf{W} \mathbf{\Phi} (\mathbf{\Phi}^T \mathbf{W} \mathbf{\Phi})^{-1}.$$

Using

$$\text{var}\,(\mathbf{Y}) = \mathbf{V} = \sigma^2 \mathbf{W}^{-1},$$

we have

$$\text{var}\,(\hat{\mathbf{\Theta}}) = \sigma^2 (\mathbf{\Phi}^T \mathbf{W} \mathbf{\Phi})^{-1}. \tag{8.25}$$

This is the variance matrix of the parameters. Unlike the estimation problem itself (*cf.* Eqn (8.13)), to calculate the variance matrix for $\mathbf{\Theta}$ requires a knowledge of σ^2, the scale factor in the variance matrix of the observations. Fortunately, the least-squares method allows us to calculate an unbiased estimate of σ^2 which may be used in (8.25).

To estimate σ^2 we return to Eqn (8.10), and consider the expected value of the weighted sum of residuals S.

$$\sigma^2 E[S] = E[\mathbf{R}^T\mathbf{W}\mathbf{R}]. \tag{8.26}$$

When $\mathbf{\Theta} = \hat{\mathbf{\Theta}}$ the r.h.s. of (8.26) becomes

$$E[\mathbf{R}^T\mathbf{W}(\mathbf{Y} - \mathbf{\Phi}\hat{\mathbf{\Theta}})] = E[\mathbf{R}^T\mathbf{W}\mathbf{Y}],$$

since

$$\mathbf{R}^T\mathbf{W}\mathbf{\Phi}\hat{\mathbf{\Theta}} = 0,$$

is equivalent to a statement of the normal equations. Furthermore,

$$\mathbf{R}^T\mathbf{W}\mathbf{Y} = (\mathbf{Y}^T - \hat{\mathbf{\Theta}}^T\mathbf{\Phi}^T)\mathbf{W}\mathbf{Y} = (\mathbf{Y}^T\mathbf{W}\mathbf{Y}) - (\hat{\mathbf{\Theta}}^T\mathbf{N}\hat{\mathbf{\Theta}}), \tag{8.27}$$

where

$$\mathbf{N} = \mathbf{\Phi}^T\mathbf{W}\mathbf{\Phi}.$$

By using the normal equations once again Eqn (8.27) may be reduced to

$$(\mathbf{Y} - \mathbf{Y}^0)^T\mathbf{W}(\mathbf{Y} - \mathbf{Y}^0) - (\hat{\mathbf{\Theta}} - \mathbf{\Theta})^T\mathbf{N}(\hat{\mathbf{\Theta}} - \mathbf{\Theta}),$$

and thus we have arrived at the result that

$$\begin{aligned} E[S] &= E[\mathbf{R}^T\mathbf{V}^{-1}\mathbf{R}] \\ &= E[(\mathbf{Y} - \mathbf{Y}^0)^T\mathbf{V}^{-1}(\mathbf{Y} - \mathbf{Y}^0) - (\hat{\mathbf{\Theta}} - \mathbf{\Theta})^T\mathbf{M}^{-1}(\mathbf{\Theta} - \mathbf{\Theta})], \end{aligned} \tag{8.28}$$

where

$$\mathbf{M} = \sigma^2\mathbf{N}^{-1},$$

which, by Eqn (8.25), is the variance matrix of the parameters.

Consider the first term in (8.28). The quantity $(\mathbf{Y} - \mathbf{Y}^0)$ is a vector of random variables distributed with mean zero, and with variance matrix $\mathbf{V}$. Thus

$$\begin{aligned} &E[\mathbf{Y} - \mathbf{Y}^0)^T\mathbf{V}^{-1}(\mathbf{Y} - \mathbf{Y}^0)] \\ &\quad = E[\mathrm{Tr}\,\{(\mathbf{Y} - \mathbf{Y}^0)^T\mathbf{V}^{-1}(\mathbf{Y} - \mathbf{Y}^0)\}] \\ &\quad = \mathrm{Tr}\,\{E[(\mathbf{Y} - \mathbf{Y}^0)(\mathbf{Y} - \mathbf{Y}^0)^T\mathbf{V}^{-1}]\} \\ &\quad = \mathrm{Tr}\,(\mathbf{V}\mathbf{V}^{-1}) = n. \end{aligned}$$

Similarly, since $\mathbf{M}$ is the variance matrix of $\hat{\mathbf{\Theta}}$,

$$E[(\hat{\mathbf{\Theta}} - \mathbf{\Theta})^T \mathbf{M}^{-1} (\hat{\mathbf{\Theta}} - \mathbf{\Theta})] = p.$$

Thus, from (8.28) we have

$$E[\mathbf{R}^T \mathbf{V}^{-1} \mathbf{R}] = n - p,$$

and so an unbiased estimate for σ^2 is

$$\hat{\sigma}^2 = \frac{\mathbf{R}^T \mathbf{W} \mathbf{R}}{n - p}, \tag{8.29}$$

and consequently an unbiased estimate for the variance matrix for $\hat{\mathbf{\Theta}}$ is

$$\begin{aligned} \mathbf{E} &= \frac{\mathbf{R}^T \mathbf{W} \mathbf{R}}{n - p} \cdot (\mathbf{\Phi}^T \mathbf{W} \mathbf{\Phi})^{-1} \\ &= \frac{1}{n - p} (\mathbf{Y} - \mathbf{\Phi}\hat{\mathbf{\Theta}})^T \mathbf{W} (\mathbf{Y} - \mathbf{\Phi}\hat{\mathbf{\Theta}}) (\mathbf{\Phi}^T \mathbf{W} \mathbf{\Phi})^{-1}. \end{aligned} \tag{8.30}$$

This matrix is also known loosely as the *error matrix*, and it is common practice to quote the error on the parameter $\hat{\sigma}_i$ as

$$\Delta\hat{\theta}_i = (E_{ii})^{\frac{1}{2}}.$$

To find the error of the fitted value f we can simply use Eqn (8.6), for f, together with Eqns (8.24) and (8.25). The result, which is true for all values of x, is

$$(\Delta f)^2 \equiv \operatorname{var} f(x) = \sum_{k=1}^{p} \sum_{l=1}^{p} \phi_k(x) E_{kl} \phi_l(x), \tag{8.31}$$

which could also have been obtained from Eqn (5.44). The least-squares results may be used in a simple way to combine the results of several experiments, thereby generalizing Eqns (7.24) and (7.25). The following example will illustrate this.

EXAMPLE 8.1. An experiment measures two parameters θ_i to be $\theta_1{}^{(1)} = 1{\cdot}0$ and $\theta_2{}^{(1)} = -1{\cdot}0$ with a variance matrix

$$\mathbf{V}^{(1)} = \begin{pmatrix} 2{\cdot}0 & -1{\cdot}0 \\ -1{\cdot}0 & 1{\cdot}5 \end{pmatrix} 10^{-2}.$$

A second experiment finds a new value of θ_2 to be $\theta_2{}^{(2)} = -1{\cdot}1$ with variance 10^{-2}. We wish to combine the results of the two experiments.

The variance matrix for the two experiments is

$$\mathbf{V} = \begin{pmatrix} 2{\cdot}0 & -1{\cdot}0 & 0 \\ -1{\cdot}0 & 1{\cdot}5 & 0 \\ 0 & 0 & 1{\cdot}0 \end{pmatrix} 10^{-2}.$$

Also, in the notation used previously

$$\mathbf{\Phi}^T = \begin{pmatrix} 1 & 0 & 0 \\ 0 & 1 & 1 \end{pmatrix},$$

and

$$\mathbf{Y} = \begin{pmatrix} \theta_1^{(1)} \\ \theta_2^{(1)} \\ \theta_2^{(2)} \end{pmatrix} = \begin{pmatrix} 1{\cdot}0 \\ -1{\cdot}0 \\ -1{\cdot}1 \end{pmatrix}.$$

Thus we have

$$\mathbf{W} = \mathbf{V}^{-1} = \tfrac{1}{4} \begin{pmatrix} 3 & 2 & 0 \\ 2 & 4 & 0 \\ 0 & 0 & 4 \end{pmatrix} 10^2,$$

and

$$(\mathbf{\Phi}^T \mathbf{W} \mathbf{\Phi})^{-1} = \tfrac{1}{5} \begin{pmatrix} 8 & -2 \\ -2 & 3 \end{pmatrix} 10^{-2}.$$

Thus, from (8.13) we have

$$\hat{\mathbf{\Theta}} = \begin{pmatrix} 1{\cdot}04 \\ -1{\cdot}06 \end{pmatrix},$$

i.e.

$$\hat{\theta}_1 = 1{\cdot}04 \quad \text{and} \quad \hat{\theta}_2 = -1{\cdot}06.$$

To calculate the associated error matrix we use (8.30). From (8.29)

$$\hat{\sigma}^2 = 0{\cdot}40,$$

and hence

$$\mathbf{E} = \begin{pmatrix} 0{\cdot}64 & -0{\cdot}16 \\ -0{\cdot}16 & 0{\cdot}24 \end{pmatrix} 10^{-2}.$$

It is sometimes useful to know which linear combinations of parameter estimates have zero covariances. Since $\mathbf{E}$ is a real, symmetric matrix it can be diagonalized by a unitary matrix $\mathbf{U}$. This same matrix then transforms the parameter estimates into the required linear combination.

We will conclude this discussion of the simple linear least-squares method with some general remarks. Firstly, in the above discussion we have not specified the functions $\phi_k(x)$ except that they form a linearly independent set. If we use a simple power series for $\phi_k(x)$ then the matrix $(\mathbf{\Phi}^T \mathbf{W} \mathbf{\Phi})$ is ill-conditioned for even quite moderate values of k, and the degree of ill-conditioning increases as n becomes larger. Thus serious rounding errors can occur if $\hat{\mathbf{\Theta}}$ is calculated from Eqn (8.13). If a power series, or similar form, is dictated by the requirements of a particular model, the parameters of which one requires to estimate, then one can only hope to circumvent the problem by a judicious choice of method to invert the matrix. Such techniques can be found in books on numerical analysis. However, if all that is required is *any* form which gives an adequate fit to the data then it would clearly be advantageous to choose functions such that the matrix to be inverted is diagonal. Such functions are called *orthogonal polynomials* and their construction is discussed in Appendix B.

The second remark concerns the quality of the fit achieved by the least-squares method. For this we will have to assume a distribution for the y_i, and we will take this to be normal about f_i. In this case the weighted sum of residuals S, of Eqn (8.10), is distributed as χ^2 with $n - p$ degrees of freedom. Thus, for a fit of given order p, one can calculate the probability P_p that the expected value S_e is smaller than the observed value S_0. The order of the fit is then increased until this probability reaches any desired level. To increase p below the point where $\chi^2 \sim (n - p)$ would result in apparently better fits to the data. However, to do so would ignore the fact that the y_i are random variables, and as such contain only a limited amount of information.

Another test which is used to supplement the χ^2-test is based on the F distribution of Section 6.3. This procedure is designed to test the significance of adding an additional term in the expansion (8.6), i.e. to answer the question: is θ_k different from zero? If S_p and S_{p-1} denote the values of S for fits of order p and $p - 1$, respectively, then, from the additive property of χ^2, the quantity $(S_{p-1} - S_p)$ obeys a χ^2 distribution with one degree of freedom, and which is distributed *independently* of S_p itself. Thus, the statistic

$$F = \frac{S_{p-1} - S_p}{S_p/(n - p)},$$

obeys an F distribution with 1 and $(n - p)$ degrees of freedom. From tables of the F distribution we can now find the probability P that the observed value F_0 is greater than the expected F_e. Thus if P_p corresponds to $F_0(n-p)$ then we may assume $\theta_p = 0$ with a probability P_p of being correct. It is still possible that even though $\theta_p = 0$ higher terms are non-zero, but in this case the χ^2-test would indicate that a satisfactory fit had not yet been achieved. We will discuss these points in more detail in Chapter 11, Section 11.4, after we have considered the theory of hypothesis testing in general.

8.2 Linear least-squares with constraints

It sometimes happens in practice that one has information about *some* of the parameters to be estimated. We will generalize the discussion of Section 8.1 by considering the situation where the additional information takes the form of a set of *linear constraint equations* of the form

$$C_{lp}\theta_p = Z_l,$$

or, in matrix notation

$$\mathbf{C\Theta} = \mathbf{Z}, \tag{8.32}$$

where the rank of $\mathbf{C}$ is l. This problem can be solved if we introduce the $(l \times 1)$ vector of Lagrange multipliers $\mathbf{\Lambda}$. Then the variation function that we have to consider is

$$L = (\mathbf{R}^T\mathbf{V}^{-1}\mathbf{R}) - 2\mathbf{\Lambda}^T(\mathbf{C\Theta} - \mathbf{Z}),$$

and setting $\delta L = 0$ gives

$$\delta L = 0 = 2[-\mathbf{Y}^T\mathbf{V}^{-1}\mathbf{\Phi} + \hat{\mathbf{\Theta}}_c{}^T(\mathbf{\Phi}^T\mathbf{V}^{-1}\mathbf{\Phi}) - \mathbf{\Lambda}^T\mathbf{C}]\delta\mathbf{\Theta},$$

i.e.

$$\mathbf{\Lambda}^T\mathbf{C} = \hat{\mathbf{\Theta}}_c{}^T(\mathbf{\Phi}^T\mathbf{V}^{-1}\mathbf{\Phi}) - \mathbf{Y}^T\mathbf{V}^{-1}\mathbf{\Phi}, \tag{8.33}$$

where $\hat{\mathbf{\Theta}}_c$ is the vector of estimates under the constraints.

Now we have seen previously that

$$(\mathbf{Y}^T\mathbf{V}^{-1}\mathbf{\Phi}) = \hat{\mathbf{\Theta}}^T(\mathbf{\Phi}^T\mathbf{V}^{-1}\mathbf{\Phi}), \tag{8.34}$$

where $\hat{\mathbf{\Theta}}$ is the estimate when the constraints are removed, and using this relation in Eqn (8.33) gives

$$\mathbf{\Lambda}^T\mathbf{C} = (\hat{\mathbf{\Theta}}_c - \hat{\mathbf{\Theta}})^T(\mathbf{\Phi}^T\mathbf{V}^{-1}\mathbf{\Phi}). \tag{8.35}$$

If we set

$$\mathbf{M} = \sigma^2(\mathbf{\Phi}^T\mathbf{V}^{-1}\mathbf{\Phi}) = (\mathbf{\Phi}^T\mathbf{W}\mathbf{\Phi}),$$

then

$$\sigma^2 \mathbf{\Lambda}^T \mathbf{C}\mathbf{M}^{-1}\mathbf{C}^T = (\hat{\mathbf{\Theta}}_c - \hat{\mathbf{\Theta}})^T \mathbf{C}^T$$
$$= \mathbf{Z}^T - \hat{\mathbf{\Theta}}^T \mathbf{C}^T,$$

from which we obtain our result $\mathbf{\Lambda}^T$,

$$\sigma^2 \mathbf{\Lambda}^T = (\mathbf{Z}^T - \hat{\mathbf{\Theta}}^T \mathbf{C}^T)(\mathbf{C}\mathbf{M}^{-1}\mathbf{C}^T)^{-1}. \tag{8.36}$$

Substituting (8.36) into (8.35) and solving for $\hat{\mathbf{\Theta}}_c$ gives

$$\hat{\mathbf{\Theta}}_c{}^T = \hat{\mathbf{\Theta}}^T + (\mathbf{Z}^T - \hat{\mathbf{\Theta}}^T \mathbf{C}^T)(\mathbf{C}\mathbf{M}^{-1}\mathbf{C}^T)^{-1}\mathbf{C}\mathbf{M}^{-1}. \tag{8.37}$$

This is the solution for the least-squares estimate of $\mathbf{\Theta}$ under the constraints.

To find the variance matrix for the estimates $\hat{\mathbf{\Theta}}_c$ we have, from (8.37)

$$\operatorname{var}(\hat{\mathbf{\Theta}}_c) = \sigma^2[\mathbf{M}^{-1} - \mathbf{M}^{-1}\mathbf{C}^T(\mathbf{C}\mathbf{M}^{-1}\mathbf{C}^T)^{-1}\mathbf{C}\mathbf{M}^{-1}], \tag{8.38}$$

which, using the definition of $\mathbf{M}$, may be written in terms of the weight matrix $\mathbf{W}$.

We are now left with the problem of finding an estimate for the scale parameter σ^2. This may be done in a similar way to the unconstrained problem. Thus we consider the expected value of the weighted sum of the residues under the constraints. This is

$$E[S] = E[(\mathbf{R}^T\mathbf{V}^{-1}\mathbf{R}) + (\hat{\mathbf{\Theta}}_c - \hat{\mathbf{\Theta}})^T(\mathbf{\Phi}^T\mathbf{V}^{-1}\mathbf{\Phi})(\hat{\mathbf{\Theta}}_c - \hat{\mathbf{\Theta}})], \tag{8.39}$$

where $\mathbf{R}$ is the matrix of residuals without constraints as defined previously in Eqn (8.10). Using the technique previously used we can show that the second term has an expected value of l, where l is the rank of the constraint matrix, $\mathbf{C}$, and we have already shown that the expected value of the first term is $(n - p)$. So an unbiased estimate of σ^2 is

$$\hat{\sigma}^2 = \frac{(\mathbf{R}^T\mathbf{W}\mathbf{R}) + (\hat{\mathbf{\Theta}}_c - \hat{\mathbf{\Theta}})^T(\mathbf{\Phi}^T\mathbf{W}\mathbf{\Phi})(\hat{\mathbf{\Theta}}_c - \hat{\mathbf{\Theta}})}{n - p + l}. \tag{8.40}$$

The second term may be written in a form which is independent of $\hat{\mathbf{\Theta}}_c$ by using (8.37) for

$$(\hat{\mathbf{\Theta}}_c - \hat{\mathbf{\Theta}}).$$

This gives

$$\hat{\sigma}^2 = \frac{(\mathbf{R}^T\mathbf{W}\mathbf{R}) + (\mathbf{Z} - \mathbf{C}\hat{\mathbf{\Theta}})^T(\mathbf{C}\mathbf{M}^{-1}\mathbf{C}^T)^{-1}(\mathbf{Z} - \mathbf{C}\hat{\mathbf{\Theta}})}{n - p + l}. \tag{8.41}$$

Finally, the error matrix for the parameters $\hat{\mathbf{\Theta}}_c$ is given by

$$\mathbf{E}_c = \hat{\sigma}^2[\mathbf{M}^{-1} - \mathbf{M}^{-1}\mathbf{C}^T(\mathbf{C}\mathbf{M}^{-1}\mathbf{C}^T)^{-1}\mathbf{C}\mathbf{M}^{-1}], \tag{8.42}$$

where $\hat{\sigma}^2$ is given by (8.41).

EXAMPLE 8.2. Consider the estimation problem of Example 8.1, but now with the constraint

$$\theta_1 + \theta_2 = 0.$$

With this constraint we have

$$\mathbf{C}^T = \begin{pmatrix} 1 \\ 1 \end{pmatrix} \quad \text{and} \quad \mathbf{Z} = \mathbf{0},$$

and we have already calculated $\mathbf{M}^{-1}$ in Example 8.1. It is

$$\mathbf{M}^{-1} = \tfrac{1}{5}\begin{pmatrix} 8 & -2 \\ -2 & 3 \end{pmatrix} 10^{-2}.$$

Thus, using

$$\hat{\mathbf{\Theta}} = \begin{pmatrix} 1\cdot04 \\ -1\cdot06 \end{pmatrix},$$

as obtained in Example 8.1, direct calculation from (8.37) gives

$$\hat{\mathbf{\Theta}}_c = \begin{pmatrix} 1\cdot057 \\ -1\cdot057 \end{pmatrix}.$$

Also, using (8.41) and (8.42), we find the associated error matrix to be

$$\mathbf{E}_c = \begin{pmatrix} 0\cdot122 & -0\cdot122 \\ -0\cdot122 & 0\cdot122 \end{pmatrix} 10^{-2},$$

which is singular, as expected.

8.3 Non-linear least-squares

If the fitting functions are not linear in the parameters then the weighted sum of residuals that we have to minimize is

$$S = \frac{1}{\sigma^2}[\mathbf{Y} - \mathbf{F}(\mathbf{\Theta})]^T\mathbf{W}[\mathbf{Y} - \mathbf{F}(\mathbf{\Theta})]. \tag{8.43}$$

Differentiating S with respect to $\mathbf{\Theta}$ and setting the result to zero leads to a set of non-linear simultaneous equations, and consequently presents a difficult problem to be solved. In practice S is minimized directly by an iterative procedure, starting from some initial estimates for $\mathbf{\Theta}$, which may be suggested by the theoretical model, or in extreme situations may be little more than educated guesses. We will illustrate how such a scheme might *in principle* be applied, but will defer to later a serious discussion of practical methods.

Let the initial estimate of $\mathbf{\Theta}$ be $\mathbf{\Theta}_0$. Then, if $\mathbf{\Theta}_0$ is close enough to the "true" values $\bar{\mathbf{\Theta}}$ we may expand the quantity

$$\mathbf{Y} - \mathbf{F}(\mathbf{\Theta})$$

in a Taylor series about $\mathbf{\Theta}_0$ and keep only the first term. Thus,

$$\mathbf{\Delta}_0 \equiv \mathbf{Y} - \mathbf{F}(\mathbf{\Theta}_0) \simeq \frac{\partial \mathbf{F}(\mathbf{\Theta}_0)}{\partial \mathbf{\Theta}} \boldsymbol{\delta}_0, \tag{8.44}$$

where $\boldsymbol{\delta}_0$ is a vector of small increments of $\mathbf{\Theta}$. The problem of calculating $\boldsymbol{\delta}_0$ is now reduced to one of linear least-squares, since both $\mathbf{\Delta}_0$, and the design matrix

$$\mathbf{\Phi}_0 = \frac{\partial \mathbf{F}(\mathbf{\Theta}_0)}{\partial \mathbf{\Theta}},$$

are obtainable. Given a solution for $\boldsymbol{\delta}_0$ from the normal equation, a new approximation

$$\mathbf{F}(\mathbf{\Theta}_1) \equiv \mathbf{F}(\mathbf{\Theta}_0 + \boldsymbol{\delta}_0),$$

may be calculated. This in turn will lead to a new design matrix

$$\mathbf{\Phi}_1 \equiv \frac{\partial \mathbf{F}(\mathbf{\Theta}_1)}{\partial \mathbf{\Theta}},$$

and a new vector $\mathbf{\Delta}_1$, and hence, via the normal equations, to a new incremental vector $\boldsymbol{\delta}_1$. This linearization procedure may now be iterated until the changes in $\mathbf{\Theta}$ from one iteration to the next one are very small. At the close of the iterations the variance matrix for the parameters is again taken to be the inverse of the matrix of the normal equations.

As we have emphasised, however, the above procedure is only to illustrate a possible method of finding the minimum of S. In practice several difficulties could occur, e.g. the initial estimates $\mathbf{\Theta}_0$ could be such as to invalidate the truncation of the Taylor series at its first term. In general such a method is not sure of converging to any value, let alone to values representing a true minimum of S.

The problem of minimizing S is an example of a more general class of problems which come under the heading of "optimization of a function of several variables" and is an active field of research at present. In Appendix C we consider some of the current methods which have proved successful in practice.

9 Estimation III: Other Methods

Estimation by the method of maximum likelihood, as described in Chapter 7, is a very general technique. However, several other methods are also in common use, and may be more suitable for certain applications. We shall briefly describe some of them below.

9.1 Minimum chi-square

Consider the case in which all the values of a population fall into k mutually exclusive categories c_i $(i = 1, 2, \ldots, k)$. Let p_i denote the proportion of values falling into category c_i where

$$\sum_{i=1}^{k} p_i = 1. \tag{9.1}$$

Furthermore, in a random sample of n observations, let o_i and $e_i = np_i$ denote the *observed* and *expected* frequency in category c_i where

$$\sum_{i=1}^{k} o_i = \sum_{i=1}^{k} e_i = n. \tag{9.2}$$

Now in Section 4.6 we considered the multinomial distribution with density function

$$f(r_1, r_2, \ldots, r_{k-1}) = \frac{n!}{\prod_{i=1}^{k} r_i!} \prod_{i=1}^{k} p_i^{r_i}, \tag{9.3}$$

where r_i denotes the frequency of observations in the ith category in which the true proportion of observations is p_i $(i = 1, \ldots, k)$. We recall that the multinomial density function gives exact probabilities for any set of observed frequencies

$$r_1 = o_1; r_2 = o_2; \ldots; r_k = o_k. \tag{9.4}$$

Now each r_i is distributed binomially and we have seen in previous sections that the binomial distribution tends rapidly to a Poisson distribution with both mean and variance equal to np_i. The Poisson distribution in turn tends to a normal distribution as np_i increases. Conventionally the Poisson distribution is considered approximatley normal if $\mu \gtrsim 9$. Thus if $np_i \gtrsim 9$, r_i is approximately normally distributed with mean and variance both given by np_i. If follows, that by converting to standard measure, the statistic

$$u_i = \frac{r_i - np_i}{(np_i)^{\frac{1}{2}}}, \tag{9.5}$$

is approximately normally distributed with mean zero and unit variance. Furthermore

$$\chi^2 = \sum_{i=1}^{k} u_i^2 = \sum_{i=1}^{k} \frac{(r_i - np_i)^2}{np_i} = \sum_{i=1}^{k} \frac{(o_i - e_i)^2}{e_i}, \tag{9.6}$$

is distributed as χ^2 with $(k - 1)$ degrees of freedom.

A more common situation that arises in practice is where the generating density function is not completely specified, but instead, contains a number of unknown parameters. If the observed frequencies are used to provide estimates of the p_i, then the quantity analogous to χ^2 of Eqn (9.6) is

$$\chi'^2 = \sum_{i=1}^{k} \frac{(o_i - n\hat{p}_i)^2}{n\hat{p}_i}. \tag{9.7}$$

There now arise two questions; (1) what is the best way of estimating the p_i and (2) what is the distribution of χ'^2? There are clearly many different methods available to estimate the p_i but one which is widely used is to choose values which minimize χ'^2. This may in general be a difficult problem, and is another example of the general class of optimization problems mentioned at the end of Chapter 8, and which are discussed more fully in Appendix C. It can be shown that for a wide class of methods of estimating the p_i, including that of minimum chi-square, χ'^2 is asymptotically distributed as χ^2 with $(k - 1 - c)$ degrees of freedom where c is the number of independent parameters of the distribution used to estimate the p_i.

In general, if $\mathbf{x}_i$ is a sample of size n from a multinomial population with mean $\boldsymbol{\mu}(\boldsymbol{\theta})$ and variance matrix $\mathbf{V}(\boldsymbol{\theta})$, where $\boldsymbol{\theta}$ is to be estimated, then the value $\hat{\boldsymbol{\theta}}(\mathbf{x}_1, \ldots, \mathbf{x}_n)$ which minimizes

$$\chi^2 = \frac{1}{n}(\bar{\mathbf{x}} - \boldsymbol{\mu}(\boldsymbol{\theta}))^T[\mathbf{V}(\boldsymbol{\theta})]^{-1}(\bar{\mathbf{x}} - \boldsymbol{\mu}(\boldsymbol{\theta})),$$

i.e. the minimum $-\chi^2$ estimate of $\boldsymbol{\theta}$, is known to be consistent, asymptotically efficient, and asymptotically normal distributed if $\mathbf{x}$ is distributed like the binomial, Poisson or normal distribution (and many others).

9.2 Minimum variance

Suppose we have drawn samples from n populations each with the same mean μ but with different variances. Let the sample means be $\bar{x}_i$ and the corresponding variances be $\sigma_i^2 \equiv \sigma^2(\bar{x}_i)$. We will consider the problem of combining these samples to obtain an estimate of the population mean μ.

Since $\bar{x}_i$ is an unbiased estimate of μ, the quantity

$$\bar{x} = \sum_{i=1}^{n} a_i \bar{x}_i, \tag{9.8}$$

with

$$\sum_{i=1}^{n} a_i = 1, \tag{9.9}$$

is also an unbiased estimate, regardless of the values of the coefficients a_i, so the problem is one of selecting a suitable set of a_i. One criterion that is employed is to choose the a_i such that $\bar{x}$ has minimum variance.

Now we have

$$\operatorname{var}(\bar{x}) = \operatorname{var}\left(\sum_{i=1}^{n} a_i \bar{x}_i\right)$$

$$= \sum_{i=1}^{n} a_i^2 \operatorname{var}(\bar{x}_i) = \sum_{i=1}^{n} a_i^2 \sigma_i^2, \tag{9.10}$$

and we wish to minimize (9.10) subject to the constraint (9.9). Once again we will use the method of Lagrange multipliers. Thus, if we introduce a multiplier λ, then the variation function is

$$L = \sum_{i=1}^{n} a_i^2 \sigma_i^2 + \lambda\left(\sum_{i=1}^{n} a_i - 1\right),$$

and

$$\frac{\partial L}{\partial a_i} = 0 = 2a_i\sigma_i^2 + \lambda.$$

Thus

$$a_i = -\frac{\lambda}{2\sigma_i^2}, \tag{9.11}$$

and so, from (9.9)

$$\lambda = -2 \Big/ \sum_{j=1}^{n} (1/\sigma_j^2). \tag{9.12}$$

Substituting (9.12) in (9.11) gives the solution for the coefficient a_i,

$$a_i = \frac{1/\sigma_i^2}{\sum_{j=1}^{n} 1/\sigma_j^2},$$

and so, from (9.8)

$$\bar{x} = \frac{\sum_{i=1}^{n} (\bar{x}_i/\sigma_i^2)}{\sum_{i=1}^{n} (1/\sigma_i^2)}. \tag{9.13}$$

The variance of $\bar{x}$ may be found from (9.10). It is

$$\text{var}(\bar{x}) = 1 \Big/ \sum_{j=1}^{n} (1/\sigma_j^2). \tag{9.14}$$

The above discussion illustrates the use of the minimum variance criterion to obtain an estimate. In the example given it is interesting to note that the results obtained by the minimum variance method are identical to those obtained by the maximum likelihood method (*cf.* Example 7.6), *where the population distribution was assumed to be normal.* In general, for other population densities, the results of the two methods will differ, however.

9.3 Bayes' estimators

In Section 2.3 we introduced Bayes' Theorem and showed that to maximize the *a posteriori* probability required a knowledge of *a priori* probabilities. In general, these latter probabilities are not completely known. However, there occur cases where some partial information is available, and in these circumstances it is clearly advantageous to include it in the estimation procedure.

We will consider the case where the prior information about the parameter is such that the parameter itself can be *formally* regarded as a random variable with an associated density $p(\theta)$. The form for $p(\theta)$ could be obtained, for example, by plotting all previous estimates of θ. This will very often be found to be an approximately Gaussian form, and from the results estimates of the mean and variance of the associated normal distribution could be made. In these cases where both the usual variable and the parameter may be regarded as random variables we will denote the corresponding density as $f_R(x; \theta)$.

Before proceeding to a formal definition of Bayes' estimators we shall need a few definitions of subsidiary quantities.

Firstly, borrowing from the field of decision theory we shall consider the *loss function* $l(\hat{\theta};\theta)$ which, expressed loosely, gives the "loss" incurred by using the estimate $\hat{\theta}$ instead of the true value θ. In practice it is difficult to know what form to assume for the loss function, but a simple, common-sense, form which suggests itself is

$$l(\hat{\theta};\theta) = (\hat{\theta} - \theta)^2. \tag{9.15}$$

(In fact a loss function which is bounded by zero, as in the example we have chosen, is an example of a more general function found in decision theory, called a *risk function*). The other quantities we shall need follow directly from work of previous chapters. Thus we shall call

$$j(x_1, x_2, \dots, x_n, \theta) = f(x_1, \dots x_n|\theta)p(\theta), \tag{9.16}$$

the joint density of $x_1, \dots, x_n$, and θ, and

$$m(x_1, x_2, \dots, x_n) = \int_{-\infty}^{\infty} j(x_1, x_2, \dots, x_n, \theta)d\theta, \tag{9.17}$$

the marginal distribution of the x's. From Eqn (3.30) it then follows that the conditional distribution of θ given $x_1, \dots, x_n$ is

$$c(\theta|x_1, x_2, \dots, x_n) = \frac{j(x_1, x_2, \dots, x_n, \theta)}{m(x_1, x_2, \dots, x_n)}$$

$$= \frac{f(x_1, x_2, \dots, x_n|\theta)p(\theta)}{m(x_1, x_2, \dots, x_n)}. \tag{9.18}$$

This is called the *a posteriori* density. We may now define a Bayes' estimator.

DEFINITION 9.1. Let $x_1, \dots, x_n$ be a random sample of size n drawn from a density $f_R(x;\theta)$. Let $p(\theta)$ be the density of θ, and $f(x_1, \dots, x_n|\theta)$ be the conditional density of the x's given θ. Furthermore, let $c(\theta|x_1, \dots, x_n)$ be the conditional density of θ given the x's, and let $l(\hat{\theta};\theta)$ be the loss function. Then the *Bayes' estimator* of θ is that function defined by

$$\hat{\theta} = d(x_1, x_2, \dots, x_n),$$

which minimizes the quantity

$$B(\hat{\theta}; x_1, \dots, x_n) = \int_{-\infty}^{\infty} l(\hat{\theta};\theta)c(\theta|x_1, \dots, x_n)d\theta. \tag{9.19}$$

We will illustrate the use of this definition by an example.

EXAMPLE 9.1. Let x_i $(i = 1, 2, \ldots, n)$ be a random sample of size n drawn from a normal density with unit variance

$$f_R(x; \mu) = \frac{1}{(2\pi)^{\frac{1}{2}}} \exp\left[-\tfrac{1}{2}(x - \mu)^2\right].$$

From the definitions above

$$f(x_1, \ldots, x_n|\mu) = \frac{1}{(2\pi)^{n/2}} \exp\left[-\tfrac{1}{2}(\Sigma x_i^2 - 2\mu\Sigma x_i + n\mu^2\right].$$

Now if we assume that our information about μ takes the form of knowing that it is normally distributed, i.e.

$$p(\mu) = \frac{1}{(2\pi)^{\frac{1}{2}}} \exp(-\mu^2/2), \qquad -\infty < \mu < \infty,$$

then from (9.16) and (9.17)

$$j(x_1, \ldots, x_n, \mu) = \frac{1}{(2\pi)^{(n+1)/2}} \exp\left[-\tfrac{1}{2}(\Sigma x_i^2 + (n+1)\mu^2 - 2\mu n\bar{x})\right],$$

and

$$m(x_1, \ldots, x_n) = \frac{\exp(-\frac{1}{2}\Sigma x_i^2)}{(2\pi)^{(n+1)/2}} \int_{-\infty}^{\infty} \exp\left[-\tfrac{1}{2}((n+1)\mu^2 - 2\mu n\bar{x})\right] d\mu$$

$$= \frac{1}{(n+1)^{\frac{1}{2}}(2\pi)^{n/2}} \exp\left[-\tfrac{1}{2}\left(\Sigma x_i^2 - \frac{n^2\bar{x}^2}{n+1}\right)\right].$$

Thus, from (9.18), after some simplification, we have

$$c(\mu|x_1, \ldots, x_n) = \left(\frac{n+1}{2\pi}\right)^{\frac{1}{2}} \exp\left\{-\frac{(n+1)}{2}\left[\mu - \frac{n\bar{x}}{n+1}\right]^2\right\}. \quad (9.20)$$

If we further assume the following form for the loss function,

$$l(\hat{\mu}; \mu) = (\hat{\mu} - \mu)^2, \quad (9.21)$$

then using (9.20) and (9.21) in (9.19) gives

$$B(\hat{\mu}; x_1, \ldots, x_n)$$

$$= \left(\frac{n+1}{2\pi}\right)^{\frac{1}{2}} \int_{-\infty}^{\infty} (\hat{\mu} - \mu)^2 \exp\left\{-\frac{(n+1)}{2}\left[\mu - \frac{n\bar{x}}{n+1}\right]^2\right\} d\mu$$

$$= \hat{\mu}^2 - \frac{2\hat{\mu}\bar{x}n}{n+1} + \frac{1}{n+1} + \left(\frac{\bar{x}n}{n+1}\right)^2.$$

To minimize B we have

$$\frac{\partial B}{\partial \hat{\mu}}(\hat{\mu}; x_1, \ldots, x_n) = 0,$$

giving

$$\hat{\mu} = \frac{1}{n+1}\sum_{i=1}^{n} x_i. \tag{9.22}$$

This is the Bayes' estimator for μ.

Under very general conditions it can be shown that Bayes' estimators, independent of the assumed prior distributions, are efficient, consistent, and a function of sufficient estimators. Furthermore, Bayes' estimators tend to M.L. estimators for large samples. The disadvantage in using the method in practice is the necessity of assuming a form for both $p(\theta)$ and $l(\hat{\theta}; \theta)$.

9.4 Method of moments

In Section 3.4 (Theorem 3.2) we saw that two distributions with a common m.g.f. were equal. This provides a method for estimating the parameters of a distribution by estimating the moments of the distribution.

Let $f(x; \theta_1, \ldots, \theta_p)$ be a univariate density function with p parameters θ_i $(i = 1, 2, \ldots, p)$, and let the first p moments about the origin be

$$\begin{aligned}\mu_j' &= \int_{-\infty}^{\infty} x^j f(x; \theta_1, \ldots, \theta_p) dx, \\ &= \mu_j'(\theta_1, \ldots, \theta_p), \qquad j = 1, 2, \ldots, p.\end{aligned} \tag{9.23}$$

Let x_n be a random sample of size n drawn from the density f. The first p sample moments are given by

$$m_j' = \frac{1}{n}\sum_{i=1}^{n} x_i^j. \tag{9.24}$$

The estimators $\hat{\theta}_i$ of the parameters θ_i are obtained from the solutions of the p equations

$$m_j' = \mu_j', \qquad j = 1, 2, \ldots, p. \tag{9.25}$$

EXAMPLE 9.2. Consider the normal distribution, for which we have previously seen (Eqn (4.6))

$$\mu_1' = \mu; \qquad \mu_2' = \sigma^2 + \mu^2. \tag{9.26}$$

The sample moments are

$$m_1' = \frac{1}{n}\sum_{i=1}^{n} x_i; \qquad m_2' = \frac{1}{n}\sum_{i=1}^{n} x_i^2. \tag{9.27}$$

Applying (9.25) gives

$$\hat{\mu} = \frac{1}{n}\sum_{i=1}^{n} x_i = \bar{x}, \tag{9.28}$$

and

$$\hat{\sigma}^2 + \hat{\mu}^2 = \frac{1}{n}\sum_{i=1}^{n} x_i^2, \tag{9.29}$$

i.e.

$$\hat{\sigma}^2 = \frac{1}{n}\left[\sum_{i=1}^{n} x_i^2 - n\bar{x}^2\right] = \frac{1}{n}\Sigma(x_i - \bar{x})^2. \tag{9.30}$$

Thus, the estimators obtained by the method of moments are, for this example, the same as those obtained by the maximum likelihood method.

In some applications where the population density function is not completely known it may be advantageous to use particular linear combinations of moments. Consider, for example, a density function $f(x; \theta_1, \ldots, \theta_p)$, which is unknown, but may be expanded in the form

$$f(x; \theta_1, \ldots, \theta_p) = \sum_{j=1}^{p} \theta_j P_j(x), \tag{9.31}$$

where $P_j(x)$ is a set of orthogonal polynomials normalized such that

$$\int P_i(x) P_j(x) dx = \begin{cases} \phi_j, & i = j \\ 0, & i \neq j. \end{cases} \tag{9.32}$$

The population moments deduced from (9.31) are

$$\mu_i' = \int \sum_{j=1}^{p} \theta_j P_j(x) x^j dx. \tag{9.33}$$

However, we may also consider the linear combination of moments given by

$$\Omega_i = \int \sum_{j=1}^{p} \theta_j P_j(x) P_i(x) dx, \tag{9.34}$$

which, by (9.32), are given by

$$\Omega_i = \theta_i \phi_i. \tag{9.35}$$

The equivalent sample moments are

$$m_i = \frac{1}{n} \sum_{j=1}^{n} P_i(x_j), \tag{9.36}$$

and so, by equating the two, we have

$$\hat{\theta}_i = \frac{1}{n\phi_i} \sum_{j=1}^{n} P_i(x_j). \tag{9.37}$$

This method is useful for finding the angular distribution coefficient a_j in the expansion of, e.g. a differential cross-section, i.e.

$$\frac{d\sigma}{d\cos\theta} = \sum_j a_j P_j(\cos\theta), \tag{9.38}$$

where P_j is a Legendre function. In this case

$$\hat{a}_j = \left(\frac{2j+1}{2n}\right) \sum_{i=1}^{n} P_j(x_i), \tag{9.39}$$

and since the data do not have to be grouped this method of estimation is particularly appropriate for small samples.

Many examples exist however where the data is already grouped, e.g. in the form of a histogram, and in this case some error is introduced if the moments of the sample are calculated by assuming that the frequencies are concentrated at the mid-points of the intervals. In many cases it is possible to make corrections for this effect by a method due to Sheppard. If m' are the true moments, and $\overline{m}'$ the moments as calculated from the grouped data with interval width h, then

$$\begin{aligned}
m_1' &= \overline{m}_1' \\
m_2' &= \overline{m}_2' - \tfrac{1}{12} h^2 \\
m_3' &= \overline{m}_3' - \tfrac{1}{4} \overline{m}_1' h^2 \\
m_4' &= \overline{m}_4' - \tfrac{1}{2} \overline{m}_2' h^2 + \tfrac{7}{240} h^4,
\end{aligned}$$

and, in general

$$m_r' = \sum_{j=0}^{r} \left\{ \binom{r}{j} (2^{1-j} - 1) B_j h^j \bar{m}_{r-j}' \right\},$$

where B_j is the Bernoulli number of order j. The circumstances under which it is valid to apply these corrections are discussed in detail in the book of Kendall and Stuart.

The modifications necessary to the above simple account in order to make the method of moments practically useful are similar to those which have been discussed for the least-squares method, and so we will not discuss the method further.

Under quite general conditions, it can be shown that estimators obtained by the method of moments are consistent, but not, in general, most efficient.

10 Confidence Intervals and Regions

In Chapters 7, 8 and 9 we have discussed what is usually referred to as *point estimation*, i.e. the estimation of the value of a parameter. In practice, however, point estimation alone is not enough. It is necessary to supply also some statement about the error on the estimate. This problem, known as *interval estimation*, will be examined in this chapter.

10.1 Introduction

Our aim is to find an interval about the estimator $\hat{\theta}$ such that we may make probabilistic statements concerning the probability of the true value θ being within the interval. A method which is applicable in many cases is the following. One finds, if possible, a function of the sample data and the parameter to be estimated, say u, which has a distribution independent of the parameter. Then a probability statement of the form

$$P[u_1 \leqslant u \leqslant u_2] = p,$$

is constructed and converted into a probability statement about the parameter to be estimated. It is not always possible to find such a function, and in these cases more general methods (to be described in Section 10.3) must be used. For the present we will illustrate the above method by an example.

EXAMPLE 10.1. Consider the case of sample of size 100 drawn from a population with unit variance but unknown mean μ. The quantity

$$u = \left(\frac{\hat{\mu} - \mu}{\sigma/\sqrt{n}}\right) = 10(\hat{\mu} - \mu), \tag{10.1}$$

is, in general, normally distributed with mean zero and unit variance, and thus has a density function

$$f(u) = \frac{1}{(2\pi)^{\frac{1}{2}}} \exp\left[-u^2/2\right], \tag{10.2}$$

which is independent of μ. The probability that u lies between any two arbitrary values u_1 and u_2 is thus

$$P[u_1 \leqslant u \leqslant u_2] = \int_{u_1}^{u_2} f(t)dt. \tag{10.3}$$

For example, if $u_1 = -u_2 = -1{\cdot}96$ then, from tables of the normal distribution,

$$P[-1{\cdot}96 \leqslant u \leqslant 1{\cdot}96] = \int_{-1{\cdot}96}^{1{\cdot}96} f(t)dt = 0{\cdot}95. \tag{10.4}$$

Using (10.1) this becomes

$$P[\hat{\mu} - 0{\cdot}196 \leqslant \mu \leqslant \hat{\mu} + 0{\cdot}196] = 0{\cdot}95. \tag{10.5}$$

Now suppose $\hat{\mu}$ is estimated from the sample to be $\bar{x} = 1{\cdot}0$, then we have

$$P[0{\cdot}804 \leqslant \mu \leqslant 1{\cdot}196] = 0{\cdot}95. \tag{10.6}$$

Equation (10.6) expresses the result that the probability that the *random interval* 0·804 to 1·196 contains the true mean μ is 0·95. Thus, if samples of size 100 were repeatedly drawn from the population, and if a random interval was computed for each sample from (10.5), then 95% of those intervals would be expected to contain the true mean. The interval (0·804 to 1·196 in this case) called a 95% *confidence interval*, and the probability, here 0·95, is called the *confidence coefficient*.

The above example leads directly to the following definition.

DEFINITION 10.1. A random interval (I_1, I_2), depending only on the observed data, and having the property that $100(1 - 2\alpha)\%$ of such intervals computed will include the true value of the parameter being estimated, is called the $100(1 - 2\alpha)\%$ *confidence interval* and $(1 - 2\alpha)$ is called the *confidence coefficient*. (The reason for using the quantity $(1 - 2\alpha)$ will become clear in Chapter 11.)

It is clear from Eqn (10.3) that there exist many pairs of numbers u_1 and u_2 such that $P(u_1 \leqslant u \leqslant u_2)$ is a constant. The best confidence interval is clearly the shortest, and for symmetric distributions of the type (10.2) this condition is obtained by choosing u_1 and u_2 such that $f(u_1) = f(u_2)$. In other cases the construction of confidence intervals which are shortest for a given confidence coefficient is difficult, or may not even be possible.

EXAMPLE 10.2. A practical problem which frequently arises is to put an upper limit on the observation of a rare event. For example, if out of 1000 decays of a particle, 9 are observed to be of a type E what can one say about the branching ratio for E? The Poisson distribution is applicable here and we have $\mu = \sigma^2 = 9$. However, we also know that for $\mu \geqslant 9$ the Poisson distribution is well approximated by a normal distribution. Thus the quantity

$$u = \frac{x - \mu}{\sigma} = \frac{x - 9}{3},$$

is a standard normal deviate. Thus, for example, from the tables in Appendix D,

$$P[u \leqslant 1{\cdot}645] = 0{\cdot}95,$$

and hence $x \leqslant 13{\cdot}9$ and $B_E \leqslant 1{\cdot}4\%$ with 95% confidence. Had less than 9 events been observed then the normal approximation could not have been used. Thus, if no events had been seen, a branching ratio based on one assumed event can be found from tables of the cumulative Poisson distribution to be $B_E \leqslant 0{\cdot}3\%$ with 92% confidence.

The concept of interval estimation for a single parameter may be extended in a straightforward way to include simultaneous estimation of several parameters. Thus, a $100(1 - 2\alpha)\%$ *confidence region* is a region constructed from the sample such that, for repeatedly drawn samples, $100(1 - 2\alpha)\%$ of the regions would be expected to contain the set of parameters under estimation.

It should be remarked immediately that confidence intervals and regions are essentially arbitrary, because they depend on what function of the observations is chosen to be an estimator. This fact is easily illustrated by reference to the normal distribution of Example 10.1. If we use the sample mean as an estimator of the population mean, then for a confidence coefficient of 0·95

$$P\left[\bar{x} - \frac{1{\cdot}96\sigma}{\sqrt{n}} \leqslant \mu \leqslant \bar{x} + \frac{1{\cdot}96\sigma}{\sqrt{n}}\right] = 0{\cdot}95, \qquad (10.7)$$

and the length of the interval is $2 \times 1{\cdot}96\sigma/\sqrt{n}$. However, we could also use any given single observation to be an estimator, in which case the confidence interval would be $(n)^{\frac{1}{2}}$ times as long. An important property of M.L.E.'s is that, for large samples, they provide confidence intervals and regions which are smaller, on average, than intervals and regions determined by any other method of estimation of the parameters.

10.2 Normal distributions

Because the normal distribution is of such importance we will consider obtaining confidence intervals for its parameters separately.

10.2.1. Confidence Intervals for the Mean

From Eqn (10.7) it is clear that a confidence interval for μ cannot be calculated unless the variance is known. For large samples we could use an estimate $\hat{\sigma}^2$ for this quantity without real loss of precision, but for small samples this procedure is not satisfactory. The solution is to use the quantity

$$t = \frac{\bar{x} - \mu}{(s^2/n)^{\frac{1}{2}}} = \frac{\bar{x} - \mu}{\left[\dfrac{1}{n(n-1)} \sum_{i=1}^{n} (x_i - \bar{x})^2\right]^{\frac{1}{2}}}, \qquad (10.8)$$

which, from Theorem 6.5, has a t distribution with $(n - 1)$ degrees of freedom, and only involves μ. Thus we can find a number t_α such that

$$P[-t_\alpha \leqslant t \leqslant t_\alpha] = \int_{-t_\alpha}^{t_\alpha} f(t; n-1)dt = (1 - 2\alpha). \qquad (10.9)$$

As in Example 10.1, we may now transform the inequality in (10.9) to give

$$P[(\bar{x} - T_\alpha) \leqslant \mu \leqslant (\bar{x} + T_\alpha)] = (1 - 2\alpha), \qquad (10.10)$$

where

$$T_\alpha = t_\alpha \left[\frac{1}{n(n-1)} \sum_{i=1}^{n} (x_i - \bar{x})^2\right]^{\frac{1}{2}}.$$

The width of the interval is then $2T_\alpha$. The number t_α is called the $100\alpha\%$ *level of t*, and gives the point which cuts off $100\alpha\%$ of the area under the curve $f(t)$ on the upper tail.

10.2.2. Confidence Intervals for the Variance

To set up confidence intervals for the variance we use the χ^2 distribution. Thus the quantity

$$\chi^2 = \frac{1}{\sigma^2} \sum_{i=1}^{n} (x_i - \bar{x})^2, \qquad (10.11)$$

has, from Theorem 6.2, a χ^2 distribution with $(n - 1)$ degrees of freedom, and so we can use it to find numbers ${\chi_1}^2$ and ${\chi_2}^2$ such that

$$P[{\chi_1}^2 \leqslant \chi^2 \leqslant {\chi_2}^2] = \int_{{\chi_1}^2}^{{\chi_2}^2} f(\chi^2; n-1)d\chi^2 = 1 - 2\alpha,$$

or, equivalently,

$$P\left[\frac{1}{\chi_2^2}\sum_{i=1}^{n}(x_i-\bar{x})^2 \leqslant \sigma^2 \leqslant \frac{1}{\chi_1^2}\sum_{i=1}^{n}(x_i-\bar{x})^2\right] = 1-2\alpha. \quad (10.12)$$

Since the χ^2 distribution is not symmetric, the shortest confidence interval cannot be simply obtained for a given α. However, provided the number of degrees of freedom is not too small, a good approximation is to choose χ_1^2 and χ_2^2 such that $100\alpha\%$ of the area of $f(\chi^2)$ is cut off from each tail, i.e. such that

$$\int_{\chi_1^2}^{\infty} f(\chi^2; n-1)d\chi^2 = 1-\alpha,$$

and

$$\int_{\chi_2^2}^{\infty} f(\chi^2; n-1)d\chi^2 = \alpha.$$

Such numbers can easily be obtained from tables of the χ^2 distribution function.

10.2.3. Confidence Regions for the Mean and Variance

In constructing a confidence region for the mean and variance simultaneously we cannot use the region bounded by the limits of the confidence intervals obtained separately for μ and σ^2 (a rectangle in the (μ, σ^2) plane), i.e. Eqns (10.10) and (10.12) above, because the quantities t of (10.8) and μ are not independently distributed, and hence the joint probability that the two intervals contain the true parameter values is not equal to the product of the separate probabilities. However, the distributions of $\bar{x}$ and $\Sigma(x_i-\bar{x})^2$ *are* independent and may be used to construct the required confidence region. Thus, for a $100(1-2\alpha)\%$ confidence region we may find numbers a_i such that

$$\begin{aligned} P\left[-a_1 \leqslant \left(\frac{\bar{x}-\mu}{\sigma/\sqrt{n}}\right) \leqslant a_2\right] &= (1-2\alpha)^{\frac{1}{2}}, \\ P\left[-a_3 \leqslant \left(\frac{\Sigma(x_i-\bar{x})^2}{\sigma^2}\right) \leqslant a_4\right] &= (1-2\alpha)^{\frac{1}{2}}. \end{aligned} \quad (10.13)$$

The joint probability is then $(1-2\alpha)$ by virtue of the independence of the variables. The region defined by (10.13) will not, in general, be the smallest possible but will not differ much from the minimum (which is roughly elliptical) unless the sample size is very small.

10.3 General method

The method used in Section 10.1 requires the existence of functions of the sample and parameters which are distributed independently of the parameters. This is its disadvantage, for in many cases such functions do not exist. However, for these cases there exists a more general method which we describe below.

Let $g(\hat{\theta};\theta)$ be the density of $\hat{\theta}$, the estimator for samples of size n of the parameter θ in a population density $f(x;\theta)$. For a fixed value of θ a $100(1-2\alpha)\%$ confidence interval for $\hat{\theta}$ is constructed from the expression

$$P[h_1(\theta) \leqslant \hat{\theta} \leqslant h_2(\theta)] = \int_{h_1(\theta)}^{h_2(\theta)} g(\hat{\theta};\theta)d\hat{\theta} = 1 - 2\alpha. \tag{10.14}$$

Equation (10.14) may be rewritten as

$$P[\hat{\theta} \leqslant h_1(\theta)] = \int_{-\infty}^{h_1(\theta)} g(\hat{\theta};\theta)d\hat{\theta} = \alpha, \tag{10.15}$$

and

$$P[\hat{\theta} \geqslant h_2(\theta)] = \int_{h_2(\theta)}^{\infty} g(\hat{\theta};\theta)d\hat{\theta} = \alpha, \tag{10.16}$$

which determine the functions $h_1(\theta)$ and $h_2(\theta)$. If the equations $\hat{\theta} = h_1(\theta)$ and $\hat{\theta} = h_2(\theta)$ are plotted the diagram shown in Fig. (10.1) results. A vertical line through a particular value of θ, say $\bar{\theta}$, intersects $h_1(\theta)$ and $h_2(\theta)$ at the values $\hat{\theta}_1 = h_1(\bar{\theta})$ and $\hat{\theta}_2 = h_2(\bar{\theta})$ which are the $100(1-2\alpha)\%$ confidence limits.

To construct a confidence interval for θ we calculate an estimate from a sample of size n, say $\hat{\theta}_n$. A horizontal line through $\hat{\theta}_n$ cuts the curves at values θ_1 and θ_2 which, by construction, define the confidence limits, i.e.

$$P[\theta_1 \leqslant \theta \leqslant \theta_2] = 1 - 2\alpha. \tag{10.17}$$

To find the curves $h_1(\theta)$ and $h_2(\theta)$ may be a lengthy procedure. However, in some cases the values θ_1 and θ_2 may be obtained without knowing these curves. From (10.15) and (10.16), θ_1 and θ_2 are solutions of the equations

$$\begin{aligned} &\int_{-\infty}^{\hat{\theta}_n} g(\hat{\theta};\theta)d\hat{\theta} = \alpha, \\ &\int_{\hat{\theta}_n}^{\infty} g(\hat{\theta};\theta)d\hat{\theta} = \alpha, \end{aligned} \tag{10.18}$$

so if these equations can be solved the confidence interval results directly.

The general method given above can be extended to the case of confidence regions for the n parameters of the population $f(x; \theta_1, \ldots, \theta_n)$, i.e. that region R in the parameter space such that

$$P[\hat{\theta}_1, \hat{\theta}_2, \ldots, \hat{\theta}_n \text{ are contained in } R]$$
$$= \int_R \cdots \int g(\hat{\theta}_1, \hat{\theta}_2, \ldots, \hat{\theta}_n; \theta_1, \theta_2 \ldots, \theta_n) \prod_{i=1}^{n} d\hat{\theta}_i$$
$$= 1 - 2\alpha, \tag{10.19}$$

but the resulting formulas are very complicated, and will not be considered here.

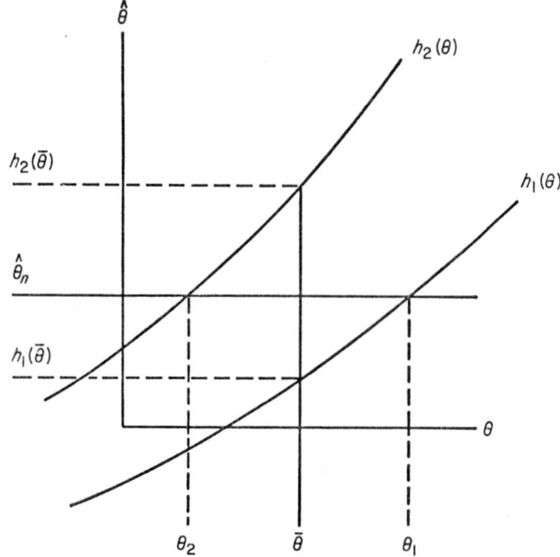

FIG. 10.1. General method for finding confidence intervals.

Finally, we note that the method cannot be used to obtain confidence regions for a subset r of the n parameters in the density $f(x; \theta_1, \ldots, \theta_n)$, except for the case of large samples which is discussed in Section 10.4. In fact this appears to be an unsolved problem for samples of arbitrary size.

10.4 Case of large samples

In Chapter 7 we have seen in Theorem 7.2 that the large sample distribution of the M.L.E. $\hat{\theta}$ of a parameter θ in the density function $f(x; \theta)$ is approximately normal about θ as mean. In this situation approximate confidence intervals may be simply constructed. The method is to convert an inequality of the form

$$P\left[-u_\alpha \leqslant \frac{\hat{\theta}-\theta}{(\operatorname{var}\hat{\theta})^{\frac{1}{2}}} \leqslant u_\alpha\right] \simeq 1-2\alpha, \tag{10.20}$$

for the distribution of $\hat{\theta}$ expressed in standard measure, to an inequality for θ itself. We recall that α is defined by

$$\frac{1}{(2\pi)^{\frac{1}{2}}}\int_{-u_\alpha}^{u_\alpha} \exp(-u^2/2)du = 1-2\alpha. \tag{10.21}$$

We will illustrate the method by applying it to the binomial distribution.

EXAMPLE 10.3. If we apply Eqn (7.27) to the binomial distribution of Eqn (4.29) we find

$$\operatorname{var}\hat{\theta} \equiv \hat{\theta}^2 = \frac{\hat{p}(1-\hat{p})}{n}. \tag{10.22}$$

From (10.20) an approximate $(1-2\alpha)$ confidence interval is obtained by considering the statement

$$P\left[-u_\alpha \leqslant \frac{\hat{p}-p}{[\hat{p}(1-\hat{p})/n]^{\frac{1}{2}}} \leqslant u_\alpha\right] \simeq 1-2\alpha, \tag{10.23}$$

which may be rewritten as

$$P\left[\hat{p}-u_\alpha\left(\frac{\hat{p}(1-\hat{p})}{n}\right)^{\frac{1}{2}} \leqslant p \leqslant \hat{p}+u_\alpha\left(\frac{\hat{p}(1-\hat{p})}{n}\right)^{\frac{1}{2}}\right] \simeq 1-2\alpha. \tag{10.24}$$

Thus, for example,

$$P\left[\hat{p}-1{\cdot}96\left(\frac{\hat{p}(1-\hat{p})}{n}\right)^{\frac{1}{2}} \leqslant p \leqslant \hat{p}+1{\cdot}96\left(\frac{\hat{p}(1-\hat{p})}{n}\right)^{\frac{1}{2}}\right] \simeq 0{\cdot}95$$

gives an approximate 95% confidence interval for p.

The above method may be extended to confidence regions by using Theorem 7.5 in place of Theorem 7.2. Thus, in terms of the matrix M_{ij} of Eqn (7.34), we know that

$$\chi^2 = \sum_{i=1}^{p}\sum_{j=1}^{p}(\hat{\theta}_i-\theta_i)M_{ij}(\hat{\theta}_j-\theta_j), \tag{10.25}$$

is approximately distributed as χ^2 with p degrees of freedom. So, just as we used (10.21) for the normal distribution, we can use the α percentage points of the χ^2 distribution to set up a confidence region for the parameters θ_i. It is an ellipsoid with centre at $(\hat{\theta}_1, \hat{\theta}_2, \ldots, \hat{\theta}_p)$.

At the end of Section 10.3 we remarked that it was not possible, in general, to obtain a confidence region for a subset of the p parameters for samples of arbitrary size. However, for large samples this *is* possible. If we wish to construct a region for a subset of r parameters ($r < p$) then the elements of the matrix M_{ij}' analogous to M_{ij} above are given by

$$M_{ij}' = (V')_{ij}^{-1}, \tag{10.26}$$

where the matrix $\mathbf{V}'$ is obtained by removing the last $(p - r)$ rows and columns in V_{ij}. The quadratic form

$$\chi'^2 = \sum_{i=1}^{r} \sum_{j=1}^{r} (\hat{\theta}_i - \theta_i) M_{ij}' (\hat{\theta}_j - \theta_j), \tag{10.27}$$

is then approximately distributed as χ^2 with r degrees of freedom, and will define an ellipsoid in the θ_i $(i = 1, \ldots, r)$ space.

11 Hypothesis Testing

11.1 Introduction

In Chapters 7–10 we have considered one of the two main branches of statistical inference, that of estimation. The other branch remaining to be discussed is that of hypothesis testing. We will start by defining what we mean by a *statistical hypothesis.*

DEFINITION 11.1. Consider a set of random variables $x_1, x_2, \ldots, x_n$, defining a sample space S of n dimensions. If we denote a general point in the sample space by E then, if R is any region in S, any hypothesis concerning the probability that E falls in R, i.e. $P(E \in R)$ is called a *statistical hypothesis.* Furthermore, if the hypothesis determines $P(E \in R)$ completely then it is called *simple*, otherwise it is called *composite.*

As an example, in testing the significance of the mean of a sample, it is a statistical hypothesis that the parent population is normal. Furthermore, if the parent population is postulated to have mean μ and variance σ^2 then the hypothesis is simple, because the density function is then completely determined.

We have already met the testing of an hypothesis when we discussed the use of the χ^2 and F distributions to determine the number of parameters needed for a satisfactory fit in the least-squares method of Chapter 8. The χ^2 distribution can also be used to test the compatibility of repeated measurements of a quantity.

EXAMPLE 11.1. If two experiments give the following results for the value of a parameter (assumed normal), $2{\cdot}05 \pm 0{\cdot}01$ and $2{\cdot}09 \pm 0{\cdot}02$. what can one say about their compatability?

The weighted mean of these results is $\bar{x} = 2{\cdot}056$ and thus $\chi^2 = 3{\cdot}25$. Since we have estimated $\bar{x}$ from the data this value of χ^2 is for one degree of freedom, and so from the tables we find

$$P(\chi^2(1) \geqslant 3{\cdot}25) \simeq 7\%.$$

Thus, a spread of this type is expected in approximately 7% of such measurements.

The above compatibility test is frequently used on data. More formally, what we are testing is the statistical hypothesis that the measurements both come from the same normal distribution.

The general procedure for testing an hypothesis is as follows. Assuming the hypothesis to be true, we can find a region R in the sample space S such that the probability of E falling in R is any preassigned value α, called the *significance level.* The region $(S - R)$ is called the *region of acceptance*, and R is called the *critical region*, or *region of rejection.* If the observed event E falls in R we reject the hypothesis, otherwise we would accept it. In practice, as we shall discuss below, the critical region is determined by a statistic, the nature of which depends upon the hypothesis to be tested. The hypothesis under test is usually called the *null* hypothesis.

By analogy with the discussion, in Chapter 10, on confidence intervals, there are many possible acceptance regions for a given hypothesis at a given significance level α. For all of them the hypothesis will be rejected, although true, in some cases. Such mistakes are called *Type I errors* and their probability, denoted by $P[\mathrm{I}]$ is equal to the significance level of the test. It is also possible that even though the hypothesis is false we fail to reject it. This is called a *Type II error.* We are led to the following definition of error probabilities.

DEFINITION 11.2. Let two hypothesis be H_1: $\theta \in \theta_1$ and H_2: $\theta \in \theta_2$, where θ_1 and θ_2 are two mutually exclusive and exhaustive regions of the parameter space. Further, let S_1 and S_2 be the acceptance and critical regions of the sample space S associated with the event $E \equiv (x_1, x_2, \ldots, x_n)$, assuming H_1 to be true. Then the probability of a Type I error is

$$P[\mathrm{I}] = P[E \in S_2 | H_1 : \theta \in \theta_1], \tag{11.1}$$

and, if H_1 is false, but H_2 is true, the probability of a Type II error is

$$P[\mathrm{II}] = P[E \in S_1 | H_2 : \theta \in \theta_2]. \tag{11.2}$$

From Eqns (11.1) and (11.2) we can define a quantity which may be used to compare the relative merits of two tests.

DEFINITION 11.3. The *power* of a statistical test is defined as

$$\begin{aligned}\beta(\theta) &\equiv P[E \in S_2 | H_2 : \theta \in \theta_2] \\ &= 1 - P[E \in S_1 | H_2 : \theta \in \theta_2].\end{aligned} \tag{11.3}$$

and is the probability of rejecting the hypothesis when it is, in fact, false. From (11.1) and (11.2) it follows that

$$\beta(\theta) = \begin{cases} P[\mathrm{I}], & \theta \in \theta_1 \\ 1 - P[\mathrm{II}], & \theta \in \theta_2. \end{cases} \tag{11.4}$$

11.2 General hypotheses: likelihood ratios

There are many techniques which have been devised to test statistical hypothesis. For example, in Section 6.1 we encountered the chi-square distribution which may be used as a "goodness-to-fit" test, i.e. may be used to test the hypothesis that a sample comes from a particular parent population. In this subsection we shall concentrate on tests based on likelihood ratios and we will start by discussing the simplest of all possible situations.

11.2.1. Simple Hypothesis: One Simple Alternative

This case is not very useful in practice but will serve as an introduction to the general methods. Firstly, we will define the *likelihood ratio test.*

Definition 11.4. A *likelihood ratio test* is used to decide between a simple null hypothesis $H_0: \theta = \theta_0$ and the simple alternative $H_a: \theta = \theta_a$. If we define the likelihood ratio λ for a sample of size n by

$$\lambda = \frac{\prod_{i=1}^{n} f(x_i; \theta_0)}{\prod_{i=1}^{n} f(x_i; \theta_a)} = \frac{L(\theta_0)}{L(\theta_a)}, \tag{11.5}$$

then, for a fixed k, the test is

for $\lambda > k$, accept H_0,

for $\lambda < k$, reject H_0,

and for $\lambda = k$, either action is taken.

The use of this test is illustrated by the following example.

Example 11.2. We will test the null hypothesis $H_0: \theta = 2$ against the alternative $H_a: \theta = 0$ for the normal population with unit variance

$$f(x; \theta) = \frac{1}{(2\pi)^{\frac{1}{2}}} \exp\left[-(x - \theta)^2/2\right].$$

From (11.5)

$$\lambda = \exp[2n\bar{x} - 2n], \tag{11.6}$$

and thus, from Definition 11.4, H_0 is accepted if

$$\lambda > k,$$

i.e. if

$$\bar{x} > c = \frac{\ln k}{2n} + 1, \tag{11.7}$$

and rejected if

$$\bar{x} < c.$$

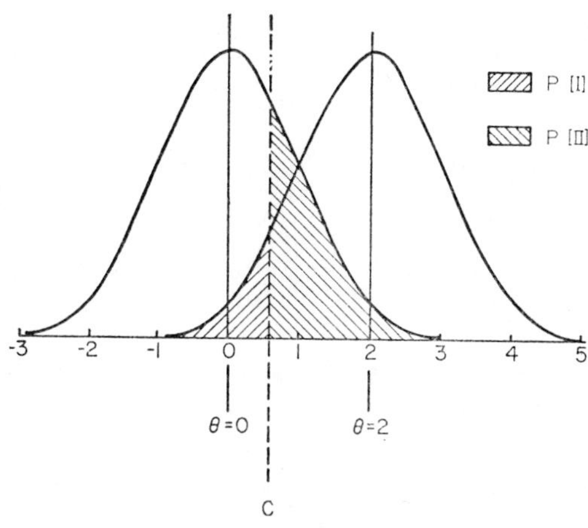

FIG. 11.1. Error probabilities for Example 11.2.

In this simple example the error probabilities are given by the shaded areas in Fig. (11.1). To find the point for which $P[\text{I}]$ is a given value for a fixed value of n we note that when $\theta = 2$

$$P[\text{I}] = P[\bar{x} < c \mid \theta = 2] = \alpha,$$

so, for $\alpha = 0{\cdot}05$ and $n = 4$ (say), using the tables in Appendix D, gives

$$c = \theta - \frac{1{\cdot}645}{\sqrt{4}} = 1{\cdot}1775,$$

and for this value of c,

$$P[\text{II}] = P[\bar{x} > 1{\cdot}1775 \mid \theta = 0] = 0{\cdot}009.$$

It is also possible to find the sample size necessary to obtain fixed values of $P[\text{I}]$ and $P[\text{II}]$. Thus, for example, if $P[\text{I}] = 0{\cdot}02$ and $P[\text{II}] = 0{\cdot}01$ we require that

$$c = \theta_0 - \frac{2{\cdot}054}{\sqrt{n}} = 2 - \frac{2{\cdot}054}{\sqrt{n}},$$

and

$$c = \theta_a + \frac{2{\cdot}323}{\sqrt{n}} = \frac{2{\cdot}323}{\sqrt{n}},$$

simultaneously. This gives $n = 4{\cdot}8$, and so a sample size of 5 would suffice.

The likelihood ratio test as used in the above example concentrates on controlling only Type I errors. A better test is one which, for a null hypotheses $H_0:\theta \in \theta_0$ with an alternative $H_a;\ \theta \in \theta_a$ gives

$$P[\text{I}] \leqslant \alpha, \quad \text{for all} \quad \theta \in \theta_0,$$

and maximizes the power

$$\beta(\theta) = 1 - P[\text{II}], \quad \text{for} \quad \text{all} \quad \theta \in \theta_a.$$

For the case of a simple null hypothesis and a simple alternative such a test is provided by the following theorem.

THEOREM 11.1. *The critical region R_k which, for a fixed significance level α, maximizes the power of the test of the null hypothesis $H_0:\theta = \theta_0$ against the alternative $H_a:\theta = \theta_a$, where $x_1, \ldots, x_n$ is a sample of size n from a density $f(x;\theta)$, is that region for which the likelihood ratio*

$$\lambda = \frac{L(\theta_0)}{L(\theta_a)} < k, \tag{11.8}$$

for a fixed number k, and

$$\int_{R_k} \cdots \int \prod_{i=1}^{n} f(x_i;\theta_0)dx_i = \alpha. \tag{11.9}$$

This result is known as the *Neyman–Pearson Lemma.*

Proof. Our object is to find the region R which maximizes the power

$$\beta = \int_R d\mathbf{x} L(\theta_a),$$

subject to the condition implied by Eqn (11.9), i.e.

$$\int_R d\mathbf{x} L(\theta_0) = \alpha.$$

Consider the region R_k defined to be that where the likelihood ratio

$$\lambda = \frac{L(\theta_0)}{L(\theta_a)} < k.$$

In R_k it follows that

$$\int_{R_k} d\mathbf{x} L(\theta_a) > \frac{1}{k} \int_{R_k} d\mathbf{x} L(\theta_0).$$

But for *all* regions Eqn (11.9) must hold, and so we have, for *any* region R

$$\int_{R_k} d\mathbf{x} L(\theta_a) > \frac{1}{k} \int_R d\mathbf{x} L(\theta_0).$$

Now for a region R outside R_k

$$\lambda = \frac{L(\theta_0)}{L(\theta_a)} > k,$$

and hence

$$\frac{1}{k} \int_R d\mathbf{x} L(\theta_0) > \int_R d\mathbf{x} L(\theta_a).$$

Combining these two inequalities gives

$$\int_{R_k} d\mathbf{x} L(\theta_a) > \int_R d\mathbf{x} L(\theta_a),$$

which is true for any R, and all R_k such that $\lambda < k$. Thus R_k is the required critical region.

We have seen in Example 11.2 that the critical region for testing the null hypothesis $H_0: \theta = 0$ against the alternative $H_a: \theta = 2$ is given by

$$\bar{x} > 1 - \frac{\ln k}{2n} = c.$$

Now if H_0 is true the distribution of x is normal with zero mean. Thus from Eqn (11.9) with $\alpha = 0{\cdot}05$,

$$\left(\frac{n}{2\pi}\right)^{\frac{1}{2}} \int_c^{\infty} \exp\left(-\frac{n\bar{x}^2}{2}\right) d\bar{x} = 0{\cdot}05,$$

giving $c = 1{\cdot}645/(n)^{\frac{1}{2}}$. Thus H_0 is to be rejected if $\bar{x} > 1{\cdot}645/(n)^{\frac{1}{2}}$.

We will illustrate the use of the Neyman–Pearson Lemma by another example, this time involving the Cauchy density of Section 4.4.

EXAMPLE 11.3. Consider the Cauchy distribution

$$f(x;\theta) = \frac{1}{\pi} \cdot \frac{1}{1 + (x - \theta)^2}\cdot$$

We will determine the critical region for a sample size one for testing the null hypothesis $H_0; \theta = 0$, against the alternative $H_a; \theta = 2$, defined by a likelihood ratio $\lambda = 5$.

Now we have, from the definition of a likelihood ratio,

$$\lambda = \frac{L(\theta = 0)}{L(\theta = 2)} = \frac{1 + (x - 2)^2}{1 + x^2}$$

and we require $\lambda < 5$. This implies that the critical regions for the test are

$$x > 0 \quad \text{and} \quad x < -1.$$

Furthermore, from Theorem 11.1, the significance level of the test is

$$\alpha = P[\text{I}] = \int_{-\infty}^{-1} dx L(\theta = 0) + \int_0^{\infty} dx L(\theta = 0)$$

$$= \frac{1}{\pi}\int_{-\infty}^{-1} dx \frac{1}{1 + x^2} + \frac{1}{\pi}\int_0^{\infty} dx \frac{1}{1 + x^2} = 0{\cdot}75.$$

Finally, we may find the probability of a Type II error. From Definition 11.2 this is

$$P[\text{II}] = \frac{1}{\pi}\int_{-1}^{0} dx \frac{1}{1 + (x - 2)^2} = 0{\cdot}045,$$

and thus the power of the test is

$$\beta = 1 - P[\text{II}] = 0{\cdot}955.$$

11.2.2. Composite Hypotheses

The cases considered in Section 11.2.1 above are really only useful for illustrative purposes. More realistic situations usually involve composite hypotheses. The first situation which we will consider is when the null hypothesis is simple and the alternative is composite, but may be regarded as an aggregate of simple hypotheses. If the alternative is H_a, then for each of the simple hypotheses in H_a, say H_a', we may construct, for a given α, a region R for testing H_0 against H_a'. However R will vary from one value of H_a' to the next and we are, therefore, faced with the problem of determining the best critical region for all the hypotheses H_a'. Such a region is called the *uniformly most powerful* (U.M.P.), and a U.M.P. *test* is defined as follows.

Definition 11.5. A test of the null hypothesis $H_0 : \theta \in \theta_0$ against the alternative $H_a : \theta \in \theta_a$, is called a *uniformly most powerful* (U.M.P.) *test* at the significance level α if the critical region of the test is such that

$$P[\text{I}] \leqslant \alpha \quad \text{for all} \quad \theta \in \theta_0,$$

$$\beta(\theta) = 1 - P[\text{II}] \quad \text{is a maximum for each} \quad \theta \in \theta_a.$$

The following simple example will illustrate how such a U.M.P. test may be constructed.

Example 11.4. We will test the null hypothesis $H_0: \mu = \mu_0$ against the alternative $H_a: \mu > \mu_0$, for a normal distribution with unit variance. The hypothesis H_a may be regarded as an aggregate of hypotheses H_a' of the form $H_a': \mu = \mu_a$ where $\mu_a > \mu_0$. The likelihood ratio for testing H_0 against H_a' is

$$\lambda = \exp\{-\tfrac{1}{2}[2n\bar{x}(\mu_a - \mu_0) + n(\mu_0^2 - \mu_a^2)]\}.$$

Theorem 11.1 may now be applied for a given k, and gives the critical region as

$$\bar{x} > c = \frac{-\ln k}{n(\mu_a - \mu_0)} + \tfrac{1}{2}(\mu_0 + \mu_a).$$

Thus the critical region is of the form $\bar{x} > c$ regardless of the value of μ_a provided $\mu_a > \mu_0$. Thus to reject H_0 if $\bar{x} > c$ tests H_0 against $H_a: \mu > \mu_0$. The number c may be found from

$$P[\text{I}] = \alpha = \left(\frac{n}{2\pi}\right)^{\frac{1}{2}} \int_c^{\infty} \exp\left[-\frac{n}{2}(\bar{x} - \mu_0)^2\right] d\bar{x},$$

by substituting $u = (n)^{\frac{1}{2}}(\bar{x} - \mu_0)$. Thus, for example, if $\alpha = 0{\cdot}05$, $c = 1{\cdot}645/(n)^{\frac{1}{2}} + \mu_0$.

A more complicated situation which can occur is in testing one composite hypothesis against another, e.g. in testing the null hypothesis $H_0: \theta_1 < \theta < \theta_2$ against $H_a: \theta < \theta_1, \theta > \theta_2$. In such cases a U.M.P. test does not exist, and other tests must be devised whose power is not too inferior to the maximum-power tests. A useful method is to construct a test having desirable large-sample properties and hope that it is still reasonable for small samples. One such test is the *generalized likelihood ratio* described below.

DEFINITION 11.6. Let $x_1, \ldots, x_n$ be a sample of size n from a population density $f(x; \theta_1, \ldots, \theta_p)$ where S is the parameter space. Let the null hypothesis be $H_0: (\theta_1, \theta_2, \ldots, \theta_p) \in R$, and the alternative be $H_a: (\theta_1, \theta_2, \ldots, \theta_p) \in (S - R)$. Then, if the likelihood of the sample is denoted by $L(S)$, and its maximum value with respect to the parameters in the region S denoted by $L(\hat{S})$, the *generalized likelihood ratio* is given by

$$\lambda = \frac{L(\hat{R})}{L(\hat{S})}, \tag{11.10}$$

and $0 < \lambda < 1$. Furthermore, if $P[\mathrm{I}] = \alpha$, then the critical region for the generalized likelihood ratio test is $0 < \lambda < A$ where

$$\int_0^A g(\lambda|H_0)d\lambda = \alpha,$$

and $g(\lambda|H_0)$ is the density of λ when the null hypothesis H_0 is true.

We will illustrate the general method by an example.

EXAMPLE 11.5. We shall test the null hypothesis $H_0: \mu = 3$ against $H_a: \mu \neq 3$, for the normal density with unit variance. In this example the region R is a single point $\mu = 3$, and $(S - R)$ is the rest of the real axis. The likelihood is

$$\begin{aligned} L &= \left(\frac{1}{2\pi}\right)^{n/2} \exp\left[-\tfrac{1}{2}\sum_{i=1}^{n}(x_i - \mu)^2\right] \\ &= \left(\frac{1}{2\pi}\right)^{n/2} \exp\left[-\tfrac{1}{2}\sum_{i=1}^{n}(x_i - \bar{x})^2 - \frac{n}{2}(\bar{x} - \mu)^2\right], \end{aligned} \tag{11.11}$$

and the maximum value of $L(S)$ is obtained when $\mu = \bar{x}$, i.e.

$$L(\hat{S}) = \left(\frac{1}{2\pi}\right)^{n/2} \exp\left[-\tfrac{1}{2}\sum_{i=1}^{n}(x_i - \bar{x})^2\right]. \tag{11.12}$$

Similarly,

$$L(\hat{R}) = \left(\frac{1}{2\pi}\right)^{n/2} \exp\left[-\tfrac{1}{2}\sum_{i=1}^{n}(x_i - \bar{x})^2 - \frac{n}{2}(\bar{x} - 3)^2\right], \tag{11.13}$$

and so the likelihood ratio is

$$\lambda = \exp\left[-\frac{n}{2}(\bar{x} - 3)^2\right]. \tag{11.14}$$

If we use $\alpha = 0{\cdot}05$, the critical region for the test is given by $0 < \lambda < A$ where

$$\int_0^A g(\lambda|H_0)d\lambda = 0{\cdot}05.$$

Now if H_0 is true, $\bar{x}$ is normally distributed with mean 3 and variance $1/n$, and $n(\bar{x} - 3)^2$ is then distributed as chi-square with one degree of freedom. From (11.14) it follows that $(-2\ln\lambda)$ is also distributed as chi-square with one degree of freedom. Setting $\chi^2 = -2\ln\lambda$, and using the tables then gives

$$0{\cdot}05 = \int_0^A g(\lambda|H_0)d\lambda = \int_{-2\ln A}^{\infty} f(\chi^2; 1)d\chi^2$$

$$= \int_{3\cdot 84}^{\infty} f(\chi^2; 1)d\chi^2.$$

Thus the critical region is defined by $-2\ln\lambda > 3{\cdot}84$, i.e.

$$(\bar{x} - 3)^2 n > 3{\cdot}84,$$

or

$$\bar{x} > 3 + \frac{1\cdot 96}{\sqrt{n}}; \qquad \bar{x} < 3 - \frac{1\cdot 96}{\sqrt{n}}. \tag{11.15}$$

We mentioned above that the generalized likelihood ratio test has useful large-sample properties. This can be stated in the following theorem.

THEOREM 11.2. *Let $x_1, x_2, \ldots, x_n$ be a random sample of size n drawn from a density $f(x; \theta_1, \ldots, \theta_p)$. Further, let the null hypothesis be*

$$H_0: \theta_i = \bar{\theta}_i, \quad i = 1, 2, \ldots, k < p,$$

with the alternative

$$H_a: \theta_i \neq \bar{\theta}_i.$$

Then, when H_0 is true, $-2\ln\lambda$ is approximately distributed as chi-square with k degrees of freedom, if n is large.

To use this theorem to test the null hypothesis H_0 with $P[\mathrm{I}] = \alpha$ we need only compute $-2\ln\lambda$ from the sample, and compare it with the α level of the chi-square distribution. If $-2\ln\lambda$ exceeds the α level H_0 is rejected, if not H_0 is accepted.

11.2.3. Fitting and Comparing Distributions

In Section 9.1 we introduced the method of estimation known as "minimum chi-square", and at the beginning of the chapter we briefly discussed how the same technique could be used to test the compatibility of repeated measurements. In the latter applications we are testing the statistical hypotheses that estimates produced by the measurement process all come from the same populations. Such procedures, for obvious reasons, are known as *goodness-of-fit tests*, and we shall discuss them further in this section.

Consider, firstly, the case of a discrete random variable x which can take on a finite number of values x_i $(i = 1, 2, \ldots, k)$ with corresponding probabilities p_i $(i = 1, 2, \ldots, k)$. We will test the null hypothesis

$$H_0: p_i = \pi_i, \quad i = 1, 2, \ldots, k,$$

against the alternative

$$H_a: p_i \neq \pi_i,$$

where the π_i are specified fixed values, and

$$\sum_{i=1}^{k} \pi_i = 1.$$

The likelihood function for a sample of size n is

$$L(\mathbf{p}) = \prod_{i=1}^{k} p_i^{\,f_i},$$

where $f_i = f_i(\mathbf{x})$ is the observed frequency of the value x_i. The maximum value of $L(\mathbf{p})$ if H_0 is true is just

$$\max L(\mathbf{p}) = L(\boldsymbol{\pi}) = \prod_{i=1}^{k} \pi_i^{\,f_i}. \tag{11.16}$$

To find the maximum value of $L(\mathbf{p})$ if H_a is true we need to know the M.L.E. of $\mathbf{p}$, i.e. $\hat{\mathbf{p}}$. Thus we have to maximize

$$\ln L(\mathbf{p}) = \sum_{i=1}^{k} f_i \ln p_i,$$

subject to the constraint

$$\sum_{i=1}^{k} p_i = 1.$$

Introducing the Lagrange multiplier Λ, the variation function becomes

$$P = \ln L(\mathbf{p}) - \Lambda \left[\sum_{i=1}^{k} p_i - 1 \right],$$

and setting

$$\frac{\partial P}{\partial p_j} = 0,$$

gives

$$p_j = f_j / \Lambda.$$

Now since

$$\sum_{j=1}^{k} p_j = 1 = \frac{1}{\Lambda} \sum_{j=1}^{k} f_j = \frac{n}{\Lambda},$$

we have that the required M.L.E. is

$$p_j = f_i / n.$$

Thus the maximum value of $L(\mathbf{p})$ if H_a is true is

$$L(\hat{\mathbf{p}}) = \prod_{i=1}^{k} \left(\frac{f_i}{n} \right)^{f_i}, \tag{11.17}$$

and the likelihood ratio is therefore, from (11.16) and (11.17),

$$\lambda = \frac{L(\boldsymbol{\pi})}{L(\hat{\mathbf{p}})} = n^n \prod_{i=1}^{k} \left(\frac{\pi_i}{f_i} \right)^{f_i}. \tag{11.18}$$

Finally, H_a is rejected if $\lambda < \lambda_c$, where λ_c is a given fixed value of λ, and H_a is accepted if $\lambda > \lambda_c$.

EXAMPLE 11.6. A die is thrown 60 times to test whether it is "true", and the resulting frequencies of the faces are as shown in the table.

Face	1	2	3	4	5	6
Frequency	9	8	12	11	6	14

Now from the above data we have, using Eqn (11.18)

$$-2\ln\lambda = -2\left[n\ln n + \sum_{i=1}^{k} f_i \ln\left(\frac{\pi_i}{f_i}\right)\right],$$

where

$$\pi_i = 1/6; \quad n = 60; \quad \text{and} \quad k = 6.$$

Thus,

$$-2\ln\lambda = -2[60\ln 60 - 60\ln 6 - 9\ln 9 \ldots - 14\ln 14]$$
$$\simeq 4{\cdot}3.$$

From Theorem 11.2 the quantity $-2\ln\lambda$ is approximately distributed as χ^2 with 5 degrees of freedom. (Note that there are only 5 degrees of freedom, not 6, because of the constraint $\Sigma p_i = 1$). From tables of the χ^2 distribution we find e.g.,

$$\chi_{0\cdot1}{}^2(5) = 9{\cdot}1,$$

and so we can certainly accept the hypothesis that the die is true, i.e.

$$H_0 : p_i = 1/6, \qquad i = 1, 2 \ldots, 6,$$

at the 10% level.

For the case of continuous distributions the null hypothesis is usually that a population is described by a certain density function $f(x)$. This hypothesis may be tested approximately by dividing the observations into k intervals and then using the method described above to compare the observed interval frequencies with those predicted by the postulated density function $f(x)$. Then, if f_i is the frequency found in the ith interval, it can be shown that

$$\chi^2 = \sum_{i=1}^{k} \frac{(f_i - n\pi_i)^2}{n\pi_i},$$

is asympototically distributed as χ^2 with $k-1$ degrees of freedom. If π_i is unknown, but is estimated from the sample in terms of r parameters to be $\hat{\pi}_i$ then the statistic

$$\chi^2 = \sum_{i=1}^{k} \frac{(f_i - n\hat{\pi}_i)^2}{n\hat{\pi}_i},$$

is also distributed as χ^2 but now with $k - 1 - r$ degrees of freedom.

In using the chi-square test of a continuous distribution with unknown parameters one has always to be careful that the method of estimating the parameters still leads to an asymptotic χ^2 distribution. In general, this will not be true if the parameters are estimated either from the original data *or* from the grouped data. The correct procedure is to estimate the parameters $\boldsymbol{\theta}$ by the M.L. method using the likelihood function

$$L(\boldsymbol{\theta}) = \prod_{i=1}^{k} [p_i(\mathbf{x}, \boldsymbol{\theta})]^{f_i},$$

where p_i is the appropriate density function. Such estimates are usually difficult to obtain, but if one uses the simple estimates then one may be working at a higher significance level than intended. This will happen, for example, in the case of the normal distribution.

11.3 Normal distribution

Because of the great importance of the normal distribution we shall give, in this section, some more details concerning tests involving this distribution.

11.3.1. Introduction

We will begin by considering the normal distribution for the situation where the population variance is known. This, of course, is not a very practical example but will serve as an introduction to more realistic cases which we will consider later.

We will thus assume that we have a normal population with density $n(x;\, \mu,\, \sigma^2)$ where the mean is unknown, but the variance *is* known. For a sample of size n drawn from the population we may compute the sample mean $\bar{x}$, and we have previously seen that

$$E[\bar{x}] = \mu, \quad \text{and} \quad \operatorname{var}(\bar{x}) = \sigma^2/n.$$

Furthermore, by the Central Limit Theorem, the variable

$$W = \left(\frac{\bar{x} - \mu}{\sigma/\sqrt{n}}\right),$$

has a density $n(W;\, 0;\, 1)$. We may now ask the question: what is the probability that $|W|$, as calculated from the sample, is greater than some specified value W_γ, where W_γ is defined by

$$P[W \geqslant W_\gamma] = \gamma. \tag{11.19}$$

Now, from the definition of the probability distribution function, we have that

$$P[|W| \geqslant W_\gamma]$$
$$= \left(\frac{1}{2\pi}\right)^{\frac{1}{2}} \left\{ \int_{-\infty}^{-W_\gamma} dt\, \mathrm{e}^{-t^2/2} + \int_{W_\gamma}^{\infty} dt\, \mathrm{e}^{-t^2/2} \right\}$$
$$= 2[1 - N(W_\gamma; 0, 1)],$$

and from Eqn (11.19) this is equal to 2γ. Substituting for W then gives the result

$$P\left[\mu - \frac{\sigma W_\gamma}{\sqrt{n}} \leqslant \bar{x} \leqslant \mu + \frac{\sigma W_\gamma}{\sqrt{n}}\right] = 2N(W_\gamma; 0, 1) - 1. \tag{11.20}$$

This inequality may either be looked upon as establishing a confidence interval for μ, or as forming the basis of a test of the hypothesis.

$$H_0\colon \mu = \mu_0; \quad f(x) = n(x; \mu, \sigma^2),$$

against the alternative

$$H_a;\ \mu \neq \mu_0; \quad f(x) = n(x; \mu, \sigma^2).$$

If we consider the latter possibility then we would be led to reject the null hypothesis if the calculated quantity

$$W_0 = \left(\frac{\bar{x} - \mu_0}{\sigma/\sqrt{n}}\right), \tag{11.21}$$

is greater than W_γ in modulus, i.e. reject H_0 if $|W_0| > W_\gamma$.

The above is a typical example of a *two-tailed test*, so-called because in such tests the probability of a Type I error

$$P[\mathrm{I}] = \alpha \equiv 2\gamma,$$

is the sum of the areas in the two tail of the normal distribution. If the alternative hypothesis was

$$H_a\colon \mu > \mu_0; \quad f(x) = n(x; \mu, \sigma^2),$$

then $P[\mathrm{I}]$ is the area under only *one* of the tails of the distribution, and the significance level of the test is thus

$$P[\mathrm{I}] = \alpha \equiv \gamma.$$

Such a test is called a *one-tailed test*.

Having fixed $P[\mathrm{I}]$ we now have to consider the probability of a Type II error. This will, of course, generally depend on the alternative hypothesis, so for definiteness we will consider the above null hypothesis with the specific alternative

$$H_a\colon \mu = \mu_a; \quad f(x) = n(x; \mu, \sigma^2).$$

Our aim is thus to find the probability that the test statistic W_0 falls within the acceptance region if the alternative hypothesis is true. We will write this as

$$P[-W_{\alpha/2} \leqslant W_0 \leqslant W_{\alpha/2} : H_a],$$

and in terms of the power, defined in Eqn (11.3),

$$P[-W_{\alpha/2} \leqslant W_0 \leqslant W_{\alpha/2} : H_a] = 1 - \beta. \tag{11.22}$$

By analogy with the definition of W_0 we shall define

$$W_a = \left(\frac{\bar{x} - \mu_a}{\sigma/\sqrt{n}}\right),$$

and if H_a is true then W_a has a normal distribution. Furthermore, since

$$W_0 = W_a + \frac{\mu_a - \mu_0}{\sigma/\sqrt{n}}, \tag{11.23}$$

we may construct an inequality for W_a by substituting (11.23) into (11.22). This gives

$$1 - \beta = P\left[-W_{\alpha/2} - \frac{\mu_a - \mu_0}{\sigma/\sqrt{n}} \leqslant W_\alpha \leqslant W_{\alpha/2} + \frac{\mu_a - \mu_0}{\sigma/\sqrt{n}}\right]. \tag{11.24}$$

Equation (11.24) shows that if $\mu_a - \mu_0$ is small then

$$P[\mathrm{II}] \simeq 1 - P[\mathrm{I}],$$

and hence

$$\beta \simeq \alpha.$$

Thus the power of the test will be very low. This situation can only be improved by making $\mu_a - \mu_0$ large, or by having n large. This is in accord with the common-sense view that it is difficult to distinguish between two close alternatives without a large quantity of data. The situation is illustrated in Fig. (11.2) where we have plotted the power β against the auxilliary parameter

$$\Delta \equiv \left(\frac{\mu_a - \mu_0}{\sigma/\sqrt{n}}\right),$$

for some sample values of α, the significance level.

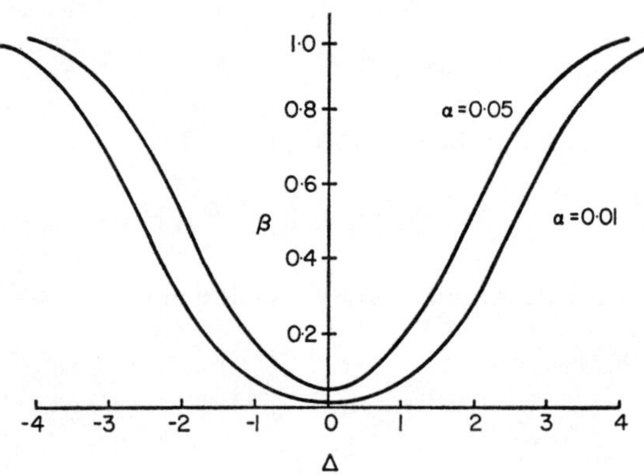

FIG. 11.2. The power of a test comparing two means for a normal population with known variance.

EXAMPLE 11.7. Four measurements of a quantity give values

$$1{\cdot}12; \quad 1{\cdot}13; \quad 1{\cdot}10; \quad 1{\cdot}09.$$

If they all come from normal populations with $\sigma^2 = 4 \times 10^{-4}$ we can apply the χ^2 goodness-of-fit test to see if it is a reasonable assumption that the populations all have the same mean, i.e. are identical. The sample mean is

$$\bar{x} = 1{\cdot}11,$$

and so

$$\chi^2 = \frac{1}{\sigma^2} \sum_{i=1}^{4} (\bar{x} - x_i)^2 = 2{\cdot}5,$$

and this value is for three degrees of freedom. From the tables we have

$$P[\chi^2(3) \geqslant 2{\cdot}5] \simeq 0{\cdot}47,$$

and so it is very reasonable to assume that the measurements all came from the same normal population. We can now test the hypothesis that the common mean is some particular value, e.g.

$$H_0 \colon \mu = \mu_0 = 1{\cdot}09,$$

against the alternative

$$H_a;\ \mu \neq 1{\cdot}09.$$

Proceeding as outlined above we calculate the test statistic

$$|W_0| = \left|\frac{\bar{x} - \mu_0}{\sigma/\sqrt{n}}\right| = 2{\cdot}0.$$

Thus, for a two-tailed test at a 5% significance level, i.e. $\alpha = 0{\cdot}05$, $W_{\alpha/2} = 1{\cdot}96$, and since

$$|W_0| > W_{\alpha/2},$$

the null hypothesis can be rejected. A 95% confidence interval for μ is then

$$\bar{x} - \frac{\sigma W_{\alpha/2}}{\sqrt{n}} \leqslant \mu \leqslant \bar{x} + \frac{\sigma W_{\alpha/2}}{\sqrt{n}},$$

i.e.

$$1{\cdot}090 \leqslant \mu \leqslant 1{\cdot}130.$$

If the alternative hypothesis was

$$H_a \colon \mu = \mu_a = 1{\cdot}11$$

then the auxiliary parameter Δ is

$$\Delta = \frac{\mu_a - \mu_0}{\sigma/\sqrt{n}} = 2$$

and so for $\alpha = 0{\cdot}05$, $\beta \simeq 0{\cdot}5$, i.e. if only the two possibilities for μ had been available, and if the test statistic fell inside the acceptance region, i.e. we accepted H_0, then the probability of having made an incorrect decision would have been $\simeq 50\%$.

We are now in a position to review the general procedure followed to test an hypothesis.

(i) State the null hypothesis H_0, and its alternative H_a.

(ii) Specify $P[\mathrm{I}]$ and $P[\mathrm{II}]$, the probabilities for errors of Types I and II, respectively, and compute the necessary sample size n. In practice

$P[\text{I}] = \alpha$ and n are commonly given. However, since even a relatively small $P[\text{II}]$ is usually of practical importance, a check should always be made to ensure that the values of α and n used lead to a suitable $P[\text{II}]$. Tables for this purpose applying to some of the test we will consider are given in Appendix D.

(iii) Choose a test statistic and determine the critical region for the test.

(iv) Accept or reject the null hypothesis H_0 depending on whether or not the value obtained for the sample statistic falls inside or outside the critical region.

A graphical interpretation of the above scheme is shown in Fig. 11.3. The curve $f_0(t|H_0)$ is the density function of the test statistic t if H_0 is true and $f_a(t|H_a)$ is its density function if H_a is true. The hypothesis H_0 is rejected if $t > t_\alpha$, and H_a is rejected if $t < t_\alpha$. The probabilities of the errors of Types I and II are also shown.

It is, perhaps, worth emphasizing that failure to reject an hypothesis does not necessarily mean that the hypothesis is true. However, if we can, on the basis of the test, reject the hypothesis then we *can* say that there is experimental evidence against it.

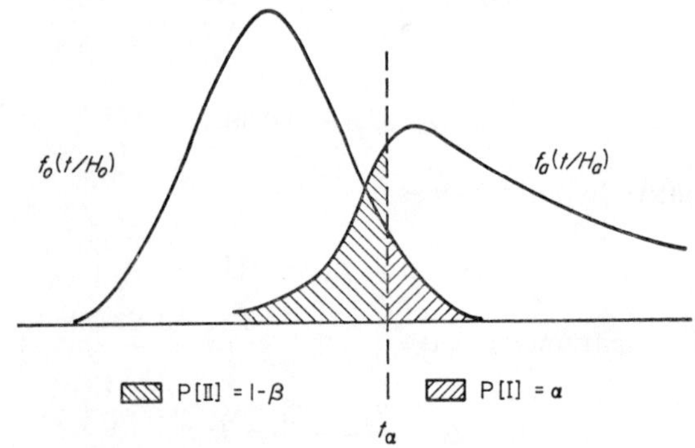

FIG. 11.3. Graphical representation of a general hypothesis test.

11.3.2. SPECIFIC TESTS

We shall now turn to the more practical case where the variance is unknown, and use the general procedure outlined at the end of Section 11.3.1 to establish some commonly used tests

(a) *Test of whether the mean is different from some specified value.*

The null hypothesis in this case is

$$H_0 : \mu = \mu_0, \quad 0 < \sigma^2 < \infty,$$

and the alternative is

$$H_a: \mu \neq \mu_0, \quad 0 < \sigma^2 < \infty.$$

The parameter space is

$$S = \{-\infty < \mu < \infty; \quad 0 < \sigma^2 < \infty\},$$

and the acceptance region associated with the null hypothesis is

$$R = \{\mu = \mu_0; \quad 0 < \sigma^2 < \infty\}.$$

The variance σ^2 is now unknown, but for large samples we could use an estimate s^2, and then take over the results of Section 11.3.1. However, for small samples this procedure could lead to gross errors and so we must devise a test where σ^2 is not explicitly used. Such a test is based on the use of the Student t distribution which was discussed in Chapter 6. We shall derive the test as follows.

The likelihood function for a sample of size n drawn from the population is given by

$$L = \frac{1}{(2\pi)^{n/2}} \cdot \frac{1}{\sigma^n} \exp\left[-\tfrac{1}{2}\sum_{i=1}^{n}\left(\frac{x_i - \mu}{\sigma}\right)^2\right]. \tag{11.25}$$

The M.L.E.'s of μ and σ^2 are

$$\hat{\mu} = \frac{1}{n}\sum_{i=1}^{n} x_i = \bar{x},$$

and

$$\hat{\sigma}^2 = \frac{1}{n}\sum_{i=1}^{n}(x_i - \bar{x})^2. \tag{11.26}$$

Using (11.26) in (11.25) gives

$$L(\hat{S}) = \left[\frac{n}{2\pi\Sigma(x_i - \bar{x})^2}\right]^{n/2} e^{-n/2}. \tag{11.27}$$

To maximize L in R we set $\mu = \mu_0$ giving

$$L' = \frac{1}{(2\pi)^{n/2}} \frac{1}{\sigma^n} \exp\left[-\tfrac{1}{2}\sum_{i=1}^{n}\left(\frac{x_i - \mu_0}{\sigma}\right)^2\right].$$

Then the value of σ^2 which maximizes L' is

$$\hat{\sigma}^2 = \frac{1}{n} \sum_{i=1}^{n} (x_i - \mu_0)^2,$$

and hence

$$L(\hat{R}) = \left[\frac{n}{2\pi\Sigma(x_i - \mu_0)^2}\right]^{n/2} e^{-n/2}. \tag{11.28}$$

From (11.27) and (11.28) the generalized likelihood-ratio is

$$\lambda = \left\{\frac{\Sigma(x_i - \bar{x})^2}{\Sigma(x_i - \mu_0)^2}\right\}^{n/2}. \tag{11.29}$$

We must now find the distribution of λ if H_0 is true. Rewriting (11.29) gives

$$\lambda = \left[1 + \frac{n(\bar{x} - \mu_0)^2}{\Sigma(x_i - \bar{x})^2}\right]^{-n/2}$$

$$= \left(1 + \frac{t^2}{n - 1}\right)^{-n/2}, \tag{11.30}$$

where

$$t = \left[\frac{n(n-1)}{\Sigma(x_i - \bar{x})^2}\right]^{\frac{1}{2}} (\bar{x} - \mu_0),$$

is distributed as the t-distribution with $(n - 1)$ degrees of freedom. Thus, from (11.30) a critical region of the form $0 < \lambda < A$ is equivalent to the region $t^2 > F(A)$. Thus a significance level of α corresponds to the pair of intervals

$$t < -t_{\alpha/2} \quad \text{and} \quad t > t_{\alpha/2},$$

where

$$\int_{t_{\alpha/2}}^{\infty} f(t; n-1)dt = \alpha/2,$$

and $f(t; n - 1)$ is the t-distribution with $(n - 1)$ degrees of freedom. If t lies between $-t_{\alpha/2}$ and $t_{\alpha/2}$, H_0 is accepted, otherwise it is rejected. This is a typical example of a two-tailed test, and is exactly equivalent to constructing a $100(1 - 2\alpha)\%$ confidence interval for μ, and accepting H_0 if μ lies within it.

The above result may be summarised as follows:

Observations	n values of x
Significance level	α
Null hypothesis	H_0: $\mu = \mu_0$
Alternative hypothesis	H_a: $\mu \neq \mu_0$
Test statistic	$t = \left[\dfrac{n(n-1)}{\Sigma(x_i - \bar{x})^2}\right]^{\frac{1}{2}} (\bar{x} - \mu_0)$,

obeys a t-distribution with $n - 1$ degrees of freedom if null hypothesis is true.

Decision criterion. If observed value of t lies between $-t_{\alpha/2}$ and $t_{\alpha/2}$ where

$$\int_{t_{\alpha/2}}^{\infty} f(t; n-1)dt = \alpha/2$$

accept null hypothesis, otherwise reject it.

The above may be generalized in an obvious way to test the null hypothesis against H_a: $\mu > \mu_0$ and H_a: $\mu < \mu_0$. The test statistic is the same, but the critical regions are now $t < t_\alpha$ and $t > t_{1-\alpha}$, respectively.

The above procedure, by specifying the significance level, has controlled Type I errors. We must now consider the power of the test. This is no longer a simple problem because if H_0 is not true then the statistic t no longer obeys a Student t distribution. However, if the alternative hypothesis is

$$H_a\colon \mu = \mu_a, \qquad 0 < \sigma^2 < \infty,$$

then, if H_a is true, it can be shown that t obeys a *non-central t-distribution* of the form

$$f^{nc}(t) = \frac{2^{-(\nu-1)/2}}{\Gamma(\nu/2)(\nu\pi)^{\frac{1}{2}}}\left(1 + \frac{t^2}{\nu}\right)^{-(\nu+1)/2} \exp\left[-\tfrac{1}{2}\left(\frac{\delta^2}{1 + t^2/\nu}\right)\right]$$

$$\times \int_0^{\infty} dx\, x^{\nu} \exp\left[-\tfrac{1}{2}\left(x - \frac{t\nu}{(t^2+\nu)^{\frac{1}{2}}}\right)\right],$$

where

$$\nu = n - 1,$$

and

$$\delta = \frac{|\mu_a - \mu_0|}{\sigma/\sqrt{n}}.$$

Unfortunately, this distribution (apart from being very complex!) contains the population variance σ^2 in the *non-centrality parameter* δ. An *estimate* of the power of the test may be obtained by replacing σ^2 by the sample variance s^2 in the non-central distribution and then using tables of the distribution.

Non-central distributions arise typically if we wish to consider the power of a test, and generally are functions of a non-centrality parameter, which itself is a function of the alternative hypothesis and a population parameter which is unknown.

Another use of the student t distribution is contained in the following test, which we will state without proof.

(b) *Test of whether the means of two populations having the same variance differ.*

Observations — m values of x_1 and n values of x_2.

Significance level — α

Null hypothesis — H_0: $\mu_1 = \mu_2$

Alternative hypothesis — H_a: $\mu_1 \neq \mu_2$

Test statistic

$$t = (\bar{x}_1 - \bar{x}_2)\left\{\frac{mn/(m+n)}{\left[\sum_i (x_{1i} - \bar{x}_1)^2 + \sum_j (x_{2j} - \bar{x}_2)^2\right]\Big/(m+n-2)}\right\}^{\frac{1}{2}},$$

obeys a t-distribution with $m + n - 2$ degrees of freedom if null hypothesis is true.

Decision criterion. If observed value of t lies in the tange $t_{1-\alpha/2} < t < t_{\alpha/2}$ where

$$\int_{t_{\alpha/2}}^{\infty} f(t; m + n - 2)dt = \alpha/2, \tag{11.31}$$

accept null hypothesis, otherwise reject it.

We will now pass on to consider some tests associated with the variance of a normal population, and, by analogy with the discussion of tests involving the mean, we shall start by a test of whether the variance is equal to some specified value.

(c) *Test of whether the variance is equal to some specified value.*

The null hypothesis in this case is

$$H_0: \sigma^2 = {\sigma_0}^2, \qquad -\infty < \mu < \infty,$$

and the alternative is

$$H_a\colon \sigma^2 \neq \sigma_0{}^2, \quad -\infty < \mu < \infty.$$

The parameter space is

$$S = \{-\infty < \mu < \infty; \;\; 0 < \sigma^2 < \infty\},$$

and the acceptance region associated with the null hypothesis is

$$R = \{-\infty < \mu < \infty; \;\; \sigma^2 = \sigma_0{}^2\}.$$

The mean is not assumed to be known. The test will involve the use of the χ^2 distribution, which we discussed in Chapter 6, and we will derive it by the method of likelihood ratios.

The likelihood function for a sample of size n drawn from the population is given by

$$L = \left(\frac{1}{2\pi}\right)^{n/2} \frac{1}{\sigma^n} \exp\left[-\tfrac{1}{2}\sum_{i=1}^{n}\left(\frac{x_i - \mu}{\sigma}\right)^2\right].$$

In the acceptance region R we have

$$L = \left(\frac{1}{2\pi}\right)^{n/2} \frac{1}{\sigma_0{}^n} \exp\left[-\tfrac{1}{2}\sum_{i=1}^{n}\left(\frac{x_i - \mu}{\sigma_0}\right)^2\right].$$

This expression is a maximum when

$$\Sigma(x_i - \mu)^2$$

is a minimum, i.e. when $\mu = \bar{x}$, the arithmetic mean of the sample. Thus

$$L(\hat{R}) = \left(\frac{1}{2\pi}\right)^{n/2} \frac{1}{\sigma_0{}^n} \exp\left[-\tfrac{1}{2}\sum_{i=1}^{n}\left(\frac{x_i - \bar{x}}{\sigma_0}\right)^2\right],$$

i.e.

$$L(\hat{R}) = \left(\frac{1}{2\pi}\right)^{n/2} \frac{1}{\sigma_0} \exp\left[-\frac{(n-1)s^2}{2\sigma_0{}^2}\right],$$

where s^2 is the sample variance. To maximize L in S we have to solve the M.L. equations. The solutions have been given in Eqns (11.26) and hence

$$L(\hat{S}) = \left(\frac{1}{2\pi}\right)^{n/2} \frac{1}{s^n}\left(\frac{n}{n-1}\right)^{n/2} \exp\left(-n/2\right).$$

We may now form the generalized likelihood ratio

$$\lambda = \frac{L(\hat{R})}{L(\hat{S})} = \left[\frac{(n-1)s^2}{n\sigma_0^2}\right]^{n/2} \exp\left[\frac{n}{2} - \frac{(n-1)s^2}{2\sigma_0^2}\right],$$

from which we see that a critical region of the form $\lambda < k$ is equivalent to the region

$$k_1 < \frac{s^2}{\sigma_0^2} < k_2,$$

where k_1 and k_2 are constants depending on n and α, the significance level of the test. Now, if H_0 is true then $(n-1)s^2/\sigma_0^2$ obeys a χ^2 distribution with $(n-1)$ degrees of freedom and so, in principle, the required values of k_1 and k_2 could be found. However, a good approximation to the optimum values is obtained by choosing values of k_1 and k_2 using equal right and left tails of the distribution. Thus we are lead to the following test procedure.

Observations	n values of x
Significance level	α
Null hypothesis	H_0: $\sigma^2 = \sigma_0^2$
Alternative hypothesis	H_a: $\sigma^2 \neq \sigma_0^2$
Test statistic	

$$\chi_0^2 = \sum_{i=1}^{n} \left(\frac{x_i - \bar{x}}{\sigma_0}\right)^2 = \frac{s^2}{\sigma_0^2}(n-1),$$

obeys a χ^2-distribution with $n-1$ degrees of freedom if null hypothesis is true.

Decision criterion. If χ^2 lies in the interval

$$\chi_{1-\alpha/2}^2 < \chi^2 < \chi_{\alpha/2}^2,$$

where χ_α^2 is defined by analogy with Eqn (11.31), accept the hypothesis, otherwise reject it.

Again, this test may be simply adapted to deal with the hypotheses H_a: $\sigma^2 > \sigma_0^2$ and H_a: $\sigma^2 < \sigma_0^2$.

We now have to ask the question: what is the probability of a Type II error in this test? If the alternative hypothesis is

$$H_a\colon \sigma^2 = \sigma_a^2,$$

and if H_a is true, then the quantity

$$\chi_a^2 = \frac{s^2}{\sigma_a^2}(n-1),$$

will be distributed as χ^2 with $n-1$ degrees of freedom. Thus, from the definition of the power function, we have

$$\beta = 1 - P\left[\chi_{\alpha/2}^2(n-1) \geqslant (n-1)\frac{s^2}{\sigma_0^2} \geqslant \chi_{1-\alpha/2}^2(n-1)\right],$$

and therefore,

$$\beta = 1 - P\left[\chi_{\alpha/2}^2(n-1)\frac{\sigma_0^2}{\sigma_a^2} \geqslant (n-1)\frac{s^2}{\sigma_a^2} \geqslant \chi_{1-\alpha/2}^2\frac{\sigma_0^2}{\sigma_a^2}\right].$$

Having fixed the significance level α and the values of σ_0 and σ_a, we can now read off from tables the probability that a chi-square variate with $(n-1)$ degrees of freedom lies between the two limits in the square brackets.

The final test that we will consider concerns the equality of the variances of two normal populations.

(d) *Test of whether the variances of two populations differ.*

Observations	m values of x_1, n values of x_2
Significance level	α
Null hypothesis	H_0: $\sigma_1^2 = \sigma_2^2$
Alternative hypothesis	H_a: $\sigma_1^2 \neq \sigma_2^2$

Test statistic

$$F = \frac{(n-1)\sum_i (x_{1i} - \bar{x}_1)^2}{(m-1)\sum_j (x_{2j} - \bar{x}_2)^2} = \frac{s_1^2/(m-1)}{s_2^2/(n-1)},$$

obeys the F distribution with $m-1$ and $n-1$ degrees of freedom if H_0 is true.

Decision criterion. If the sample value of F lies in the range

$$\frac{1}{F_{\alpha/2}(n-1, m-1)} < F < F_{\alpha/2}(m-1, n-1),$$

where $F_{\alpha/2}$ is defined by analogy with Eqn (11.31), accept the hypothesis, otherwise reject it.

To calculate the power of the test, we note that $P[\text{II}]$ depends on the value of σ_1^2/σ_2^2. If the true value of this ratio is δ then, since $(m-1)s_1^2/\sigma_1^2$ for a sample from a normal population is distributed as $\chi^2(m-1)$, we find that s_1^2/s_2^2 is distributed as

$$\frac{\sigma_1^2}{\sigma_2^2} F(m-1, n-1) = \delta F(m-1, n-1).$$

Thus,

$$\beta = 1 - P\left[F_{1-\alpha/2}(m-1, n-1) \leqslant \frac{s_1^2}{s_2^2} \leqslant F_{\alpha/2}(m-1, n-1)\right],$$

is equivalent to

$$\beta = 1 - P\left[\frac{F_{1-\alpha/2}(m-1, n-1)}{\delta} \leqslant F \leqslant \frac{F_{\alpha/2}(m-1, n-1)}{\delta}\right].$$

For any given value of δ these limits may be found from tables of the F distribution. It can be shown by consulting these tables that the power of the F-test is rather small unless the ratio of variances is very large, a result which is in accordance with common-sense.

11.4 Linear hypotheses

In Section 8.1.2 we briefly mentioned the use of the χ^2 and F distributions as goodness-of-fit tests in connection with the use of the linear least-squares method of estimation. These applications were designed to test hypotheses concerning the quality of the approximation of the observations by some assumed expression linear in the parameters. We shall generalize that discussion now to consider some other hypothesis tests which can be performed using the least-squares results.

11.4.1. Introduction

We have seen in Section 8.1 that the weighted sum of residuals

$$S = \mathbf{R}^T\mathbf{V}^{-1}\mathbf{R},$$

where

$$\mathbf{R} = \mathbf{Y} - \mathbf{\Phi\Theta},$$

and $\mathbf{V}$ is the variance matrix of the observations, is distributed as χ^2 with $n - p$ degrees of freedom, where n is the number of observations and p is the total number of parameters θ_k ($k = 1, \ldots, p$). We also saw (*cf.* Eqn (8.28)) that

$$\mathbf{R}^T\mathbf{V}^{-1}\mathbf{R} = (\mathbf{Y} - \mathbf{Y}^0)^T\mathbf{V}^{-1}(\mathbf{Y} - \mathbf{Y}^0) - (\hat{\mathbf{\Theta}} - \mathbf{\Theta})^T\mathbf{M}^{-1}(\hat{\mathbf{\Theta}} - \mathbf{\Theta}), \quad (11.32)$$

where $\mathbf{M}$ is the variance matrix of the parameters. It follows from the additive property of χ^2 that since

$$(\mathbf{Y} - \mathbf{Y}^0)^T\mathbf{V}^{-1}(\mathbf{Y} - \mathbf{Y}^0)$$

is distributed as χ^2 with n degrees of freedom, the quantity

$$(\hat{\boldsymbol{\Theta}} - \boldsymbol{\Theta})^T\mathbf{M}^{-1}(\hat{\boldsymbol{\Theta}} - \boldsymbol{\Theta}),$$

is distributed as χ^2 with p degrees of freedom. Now to test deviations from the best least-squares values for the parameters we need to know the distribution of

$$(\hat{\boldsymbol{\Theta}} - \boldsymbol{\Theta})^T\mathbf{E}^{-1}(\hat{\boldsymbol{\Theta}} - \boldsymbol{\Theta}),$$

where $\mathbf{E}$ is the error matrix of Eqn (8.30). In the notation of Section 8.1.2

$$\mathbf{E} = \hat{\sigma}^2(\boldsymbol{\Phi}^T\mathbf{W}\boldsymbol{\Phi})^{-1},$$

and so

$$\begin{aligned}&(\hat{\boldsymbol{\Theta}} - \boldsymbol{\Theta})^T\mathbf{E}^{-1}(\hat{\boldsymbol{\Theta}} - \boldsymbol{\Theta})\\ &\quad= \frac{1}{\hat{\sigma}^2}(\hat{\boldsymbol{\Theta}} - \boldsymbol{\Theta})^T(\boldsymbol{\Phi}^T\mathbf{W}\boldsymbol{\Phi})(\hat{\boldsymbol{\Theta}} - \boldsymbol{\Theta}),\end{aligned}$$

i.e.

$$(\hat{\boldsymbol{\Theta}} - \boldsymbol{\Theta})^T\mathbf{E}^{-1}(\hat{\boldsymbol{\Theta}} - \boldsymbol{\Theta}) = (\hat{\boldsymbol{\Theta}} - \boldsymbol{\Theta})^T\mathbf{M}^{-1}(\hat{\boldsymbol{\Theta}} - \boldsymbol{\Theta})\left(\frac{\sigma^2}{\hat{\sigma}^2}\right).$$

But we have seen above that

$$(\hat{\boldsymbol{\Theta}} - \boldsymbol{\Theta})^T\mathbf{M}^{-1}(\hat{\boldsymbol{\Theta}} - \boldsymbol{\Theta}),$$

is distributed as χ^2 with p-degrees of freedom, and so

$$\frac{1}{p}(\hat{\boldsymbol{\Theta}} - \boldsymbol{\Theta})^T\mathbf{E}^{-1}(\hat{\boldsymbol{\Theta}} - \boldsymbol{\Theta}),$$

is distributed as

$$\frac{\chi^2(p)/p}{\chi^2(n-p)/(n-p)} \equiv F(p, n-p).$$

Thus to test the hypothesis

$$H_0\colon \boldsymbol{\Theta} = \boldsymbol{\Theta}_0,$$

we compute the test statistic

$$F_0 = \frac{1}{p}(\hat{\mathbf{\Theta}} - \mathbf{\Theta}_0)^T \mathbf{E}^{-1}(\hat{\mathbf{\Theta}} - \mathbf{\Theta}_0),$$

and reject the hypothesis at a significance level of α if

$$F_0 > F_\alpha(p, n-p).$$

11.4.2. General Theory

In Section 8.2, we consider the least-squares method in the presence of linear constraints on the parameters. By analogy we will now generalize the discussion in Section 11.4.1 above to include the general linear hypothesis

$$H_0\colon C_{lp}\theta_p = Z_l, \qquad l \leqslant p \tag{11.33}$$

which may be an hypothesis about all of the parameters, or any subset of them.

The null hypothesis H_0 may be tested by comparing the least-squares solution for the weighted sum of residuals when H_0 is true, i.e. S_C, with the sum in the unconstrained situation, i.e. S. In the notation of Section 8.2., the additional sum of residuals $S_A = S_C - S$, which is present if the hypothesis H_0 is true, is distributed as χ^2 with l degrees of freedom, independently of S, which itself is distributed as χ^2 with $(n-p)$ degrees of freedom. Thus the ratio

$$F = \frac{S_A/l}{S/(n-p)},$$

is distributed as $F(l, n-p)$. Using the results of Section 8.2 we can then show that

$$F = \frac{1}{l}(\mathbf{Z} - \mathbf{C}\hat{\mathbf{\Theta}})^T(\mathbf{C}\mathbf{E}\mathbf{C}^T)^{-1}(\mathbf{Z} - \mathbf{C}\hat{\mathbf{\Theta}}). \tag{11.34}$$

Thus H_0 is rejected at the α significance level if $F > F_\alpha(l, n-p)$, (compare the discussion at the end of Section 8.1.2).

Example 11.8. An experiment results in the following estimates for three parameters, based on ten measurements

$$\hat{\theta}_1 = 2; \quad \hat{\theta}_2 = 4; \quad \hat{\theta}_3 = 1,$$

with an associated error matrix

$$\mathbf{E} = \begin{pmatrix} 1 & 0 & 0 \\ 0 & 2 & 1 \\ 0 & 1 & 1 \end{pmatrix}$$

Test the hypothesis

$$H_0: \theta_1 = 0, \quad \theta_2 = 0,$$

at the 5% significance level.

We have, for the above hypothesis,

$$\mathbf{C} = \begin{pmatrix} 1 & 0 & 0 \\ 0 & 1 & 0 \end{pmatrix}; \quad \mathbf{Z} = \begin{pmatrix} 0 \\ 0 \end{pmatrix},$$

and thus the calculated value of F from Eqn (11.34) is 6. From a set of tables we can find that

$$F_\alpha(l, n - p) \equiv F_{0\cdot05}(2, 7) = 4\cdot74,$$

and so we can reject the hypothesis at this significance level.

Finally, we have to consider the power of the test of the general linear hypothesis, i.e. we have to find the distribution of F if H_0 is not true. Now $S/(n-p)$ is distributed as $\chi^2/(n-p)$ regardless of whether H_0 is true or false, but S_A/l is *only* distributed as χ^2/l if H_0 *is* true. If H_0 is false then S_A/l will in general be distributed as *non-central* χ^2 which has, for l degrees of freedom, a density function

$$f^{nc}(\chi^2; l) = \sum_{p=0}^{\infty} \left(\frac{e^{-\lambda}\lambda^p}{p!} \right) f(\chi^2; l + 2p),$$

where $f(\chi^2\ l + 2p)$ is the density function for a χ^2 variable with $(l + 2p)$ degrees of freedom, and the non-centrality parameter is

$$\lambda = \tfrac{1}{2}(\mathbf{C\Theta} - \mathbf{Z})^T(\mathbf{CMC}^T)^{-1}(\mathbf{C\Theta} - \mathbf{Z}).$$

It follows that F is distributed as a *non-central F distribution*. Tables of the latter distribution are available to construct the power curves. A feature of the non-central F distribution is that the power of the test increases as λ increases.

Appendix A

Miscellaneous Mathematics

1 Matrix algebra

A *matrix* is a two-dimensional array of numbers (taken to be real in these notes) which is written as

$$\mathbf{A} = \begin{pmatrix} a_{11} & a_{12} & \dots & a_{1n} \\ a_{21} & a_{22} & \dots & a_{2n} \\ \vdots & \vdots & & \vdots \\ a_{m1} & a_{m2} & \dots & a_{mn} \end{pmatrix},$$

where the general element in the ith row and jth column is denoted by a_{ij}. A matrix with m rows and n columns is said to be of *order* $(m \times n)$.

The *transpose* of the matrix $\mathbf{A}$ is obtained by interchanging the rows and columns of $\mathbf{A}$, and is denoted by $\mathbf{A}^T$.

Two matrices may be added if they contain the same number of rows and columns, and such addition is both commutative and associative. The (inner) product of two matrices is defined only if the number of columns in the first matrix is equal to the number of rows in the second. Thus the product

$$\mathbf{A} = \mathbf{BC},$$

requires that

$$a_{ij} = \sum_k b_{ik} c_{kj}.$$

Matrix multiplication is *not*, in general, commutative, but *is* associative. Before we can discuss division of matrices we must consider the special properties of *square* matrices, i.e. those having an equal number of rows and columns.

The *determinant* of a square $(n \times n)$ matrix $\mathbf{A}$ is defined as

$$\det \mathbf{A} \equiv |\mathbf{A}| = \Sigma\,(\pm a_{1i} a_{2j} \dots a_{nk}), \tag{A1}$$

where the summation is taken over all permutations of $i, j, \ldots, k$ where these indices are the integers 1, 2, ..., n. The positive sign is used for even permutations and the negative sign for odd permutations.

The *minor* m_{ij} of the element a_{ij} is defined as the determinant obtained from $|\mathbf{A}|$ by deleting the ith row and the jth column, and the *cofactor* of a_{ij} is defined as $(-1)^{i+j}$ times the minor m_{ij}. The *adjoint* matrix is then defined as the transposed matrix of cofactors and is denoted by $\mathbf{A}^\dagger$. Thus, if

$$\mathbf{A} = \begin{pmatrix} 2 & 1 & 3 \\ 1 & 2 & 4 \\ 2 & 1 & 1 \end{pmatrix}; \qquad \mathbf{A}^\dagger = \begin{pmatrix} -2 & 2 & -2 \\ 5 & -4 & -3 \\ -3 & 0 & 3 \end{pmatrix}.$$

If $\mathbf{A}^\dagger = \mathbf{A}^T$ the matrix A is said to be *orthogonal.*

If we form all possible square submatrices of the matrix $\mathbf{A}$ (not necessarily square), and find that at least one determinant of order r is non-zero, but that all determinants of order $(r + 1)$ vanish, then the matrix is said to be of *rank* r. A square matrix of order n with rank $r < n$ has $\det \mathbf{A} = 0$ and is said to be *singular*.

If the square matrix $\mathbf{A}$ has elements such that $a_{ij} = a_{ji}$ it is said to be *symmetric*. A particular example of a symmetric matrix is the *unit-matrix* $\mathbf{1}$ with elements equal to unity for $i = j$, and zero otherwise.

If the square matrix $\mathbf{A}$, of order n, has rank $r = n$ then it is non-singular and there exists a matrix $\mathbf{B} = \mathbf{A}^{-1}$, known as the *inverse* matrix, such that

$$\mathbf{A}\mathbf{A}^{-1} = \mathbf{A}^{-1}\mathbf{A} = \mathbf{1}.$$

This is clearly the equivalent process to division in scalar algebra. The inverse is given by

$$\mathbf{A}^{-1} = \mathbf{A}^\dagger\,|\mathbf{A}|^{-1}. \tag{A2}$$

A few further definitions will be necessary. A square matrix with elements $a_{ij} \neq 0$ for $i = j$ only is called *diagonal*, and the unit matrix given above is an example of such a matrix. The line on which the elements are non-zero is called the *principal diagonal.*

A symmetric matrix $\mathbf{A}$ is said to be *positive definite* if for any vector $\mathbf{V}$, (i) $\mathbf{V}^T\mathbf{A}\mathbf{V} \geqslant 0$ and (ii) $\mathbf{V}^T\mathbf{A}\mathbf{V} = 0$ implies $\mathbf{V} = \mathbf{0}$ where $\mathbf{0}$ is the *null* vector, i.e. a vector consisting of all zeros.

Finally, for products of matrices,

$$(\mathbf{ABC} \ldots \mathbf{D})^T = \mathbf{D}^T \ldots \mathbf{C}^T\mathbf{B}^T\mathbf{A}^T,$$

and, if $\mathbf{A}, \mathbf{B}, \mathbf{C}, \ldots, \mathbf{D}$ are all square non-singular matrices,

$$(\mathbf{ABC} \ldots \mathbf{D})^{-1} = \mathbf{D}^{-1} \ldots \mathbf{C}^{-1}\mathbf{B}^{-1}\mathbf{A}^{-1}.$$

A point worth remarking is that in practice Eqns (A1) and (A2) are rarely useful for the practical evaluation of the determinant and inverse of a matrix. For example, in the least-squares method where the matrix of the normal equations (which is positive definite), has to be inverted, the most efficient methods in common use are those based either on Choleski's decomposition of a positive definite matrix, or on Golub's factorization by orthogonal matrices, the details of which may be found in any modern textbook on numerical methods.

Matrices are frequently used in Chapter 8 to write the set of n linear equations in p unknowns

$$\sum_{j=1}^{p} a_{ij}x_j = b_i, \qquad i = 1, 2, \ldots, n$$

in the compact form

$$\mathbf{AX} = \mathbf{B},$$

where $\mathbf{A}$ is of order $(n \times p)$, $\mathbf{X}$ is an $(p \times 1)$ column vector and $\mathbf{B}$ is a $(n \times 1)$ column vector.

The set of n column vectors $\mathbf{X}_i$ $(i = 1, 2, \ldots, n)$ all of the same order, are said to be *linearly dependent* if there exist n scalars α_i $(i = 1, 2, \ldots, n)$, not all zero, such that

$$\sum_{i=1}^{n} \alpha_i \mathbf{X}_i = \mathbf{0}.$$

If no such set of scalars exists, then the set of vectors is said to be *linearly independent*. It follows that if the vectors are the columns of a square matrix $\mathbf{S}$, then if $|\mathbf{S}| \neq 0$ the columns of $\mathbf{S}$ are linearly independent. The rank of a matrix, defined previously, may thus be expressed as the greatest number of linearly independent rows or columns existing in the matrix, and so, for example, a non-singular square matrix of order $(n \times n)$ must have rank n.

2 Classical theory of minima

If $f(x)$ is a function of the single variable x which in a certain interval possesses continuous derivatives $f^{(i)}(x)$ $(i = 1, 2, \ldots, n + 1)$, then *Taylor's Theorem* states that if x and $(x + h)$ belong to this interval then

$$f(x + h) = \sum_{j=0}^{n} \frac{h^j}{j!} f^{(j)}(x) + R_n,$$

where $f^{(0)}(x) = f(x)$, and the remainder term is given by

$$R_n = \frac{h^{n+1}}{(n+1)!} f^{(n+1)}(x + \theta h), \qquad 0 < \theta < 1.$$

For a function of p-variables Taylor's expansion becomes

$$f(\mathbf{x} + t\mathbf{h}) = \sum_{j=0}^{n} \frac{t^j}{j!} (\mathbf{h}\nabla)^j f(\mathbf{x}) + R_n,$$

where $\mathbf{h}$ is the row vector $(h_1 h_2 \dots h_p)$, ∇^T is the row vector

$$\left(\frac{\partial}{\partial x_1} \frac{\partial}{\partial x_2} \cdots \frac{\partial}{\partial x_p}\right)$$

and

$$R_n = \frac{t^{n+1}}{(n+1)!} (\mathbf{h}\nabla)^{n+1} f(\mathbf{x} + \theta t\mathbf{h}), \qquad 0 < \theta < 1.$$

A *necessary* condition for a turning point (maximum, minimum or saddle point) of $f(x)$ to exist is that

$$\frac{\partial f(\mathbf{x})}{\partial x_i} = 0$$

for all $i = 1, 2, \dots, p$. A *sufficient* condition for this point to be a minimum is that the second partial derivatives exist, and that $D_i > 0$ for all $i = 1, 2, \dots, p$ where

$$D_i = \begin{vmatrix} \dfrac{\partial^2 f}{\partial x_1{}^2} & \dfrac{\partial^2 f}{\partial x_1 \partial x_2} & \cdots & \dfrac{\partial^2 f}{\partial x_1 \partial x_i} \\ \dfrac{\partial^2 f}{\partial x_2 \partial x_1} & \dfrac{\partial^2 f}{\partial x_2{}^2} & \cdots & \dfrac{\partial^2 f}{\partial x_2 \partial x_i} \\ \vdots & \vdots & & \vdots \\ \dfrac{\partial^2 f}{\partial x_i \partial x_1} & \dfrac{\partial^2 f}{\partial x_i \partial x_2} & \cdots & \dfrac{\partial^2 f}{\partial x_i{}^2} \end{vmatrix}.$$

If we seek a minimum of $f(x)$ subject to the s *equality constraints*

$$e_j(\mathbf{x}) = 0, \qquad j = 1, 2, \dots, s,$$

then the quantity to consider is the *Lagrangian form*

$$L(\mathbf{x}, \boldsymbol{\lambda}) = f(\mathbf{x}) + \sum_{j=1}^{s} \lambda_j e_j(\mathbf{x}),$$

where the constants λ_j are the so-called *Lagrange multipliers.* If the first partial derivatives of $e_j(\mathbf{x})$ exist then the required minimum is the unconstrained solution of the equations

$$e_j(\mathbf{x}) = 0, \qquad j = 1, 2, \ldots, s,$$

and

$$\frac{\partial f(\mathbf{x})}{\partial x_i} + \sum_{j=1}^{s} \lambda_j \frac{\partial e_j(\mathbf{x})}{\partial x_i} = 0, \qquad i = 1, 2, \ldots, p.$$

Appendix B

Orthogonal Polynomials

In Chapter 8, on the linear least-squares method of estimation, we encountered the problem of fitting a set of n observations y_j ($j = 1, 2, \ldots, n$) in terms of p parameters θ_k ($k = 1, 2, \ldots, p < n$) by the series

$$f_j = \sum_{k=1}^{p} \theta_k \phi_k(x_j), \tag{B1}$$

where x_j are the points at which the observations are made. This fitting procedure was done by minimizing the weighted sum of residuals

$$S = \sum_{i=1}^{n} \sum_{j=1}^{n} r_i r_j V_{ij}^{-1}, \tag{B2}$$

where

$$r_i = y_i - f_i,$$

and $\mathbf{V}$ is the variance matrix of the observations. The solution was given by

$$\hat{\mathbf{\Theta}} = (\mathbf{\Phi}^T \mathbf{W} \mathbf{\Phi})^{-1} \mathbf{\Phi}^T \mathbf{W} \mathbf{Y}, \tag{B3}$$

where $\mathbf{W} = \sigma^2 \mathbf{V}^{-1}$ is the weight matrix of the observations $\mathbf{Y}$, σ^2 is a scale factor and

$$\mathbf{\Phi} = \begin{pmatrix} \phi_1(x_2) & \phi_1(x_1) & \ldots & \phi_p(x_1) \\ \phi_1(x_2) & \phi_2(x_2) & \ldots & \phi_p(x_2) \\ \vdots & \vdots & & \vdots \\ \phi_1(x_n) & \phi_2(x_n) & \ldots & \phi_p(x_n) \end{pmatrix}.$$

As was remarked in Chapter 8, if f_j is chosen to be a power series in x_j then the matrix

$$\mathbf{E} \equiv (\mathbf{\Phi}^T \mathbf{W} \mathbf{\Phi}), \tag{B4}$$

is ill-conditioned, and the degree of ill-conditioning increases as n becomes larger. Thus, serious rounding errors could occur if $\hat{\mathbf{\Theta}}$ is calculated from

Eqn (B3). To avoid this, the fitting functions should, if possible, be chosen so that **E** is a diagonal matrix. Such functions are called *orthogonal polynomials* and we shall discuss them briefly in this Appendix.

We will assume that the observations are uncorrelated (this is the situation frequently met in practice) and denote the diagonal elements of the weight matrix by $W(x_j)$ ($j = 1, 2, \ldots, n$). Then if we fit in terms of the polynomials $\psi_k(x)$ (for $k = 1, 2, \ldots, p$) the matrix of the normal equations is diagonal if

$$\sum_{j=1}^{n} W(x_j)\psi_r(x_j)\psi_s(x_j) = 0,$$

for $r \neq s$, and from Eqn (8.14) the least-squares estimate of $\mathbf{\Theta}$ is

$$\hat{\theta}_k = \frac{\sum_{j=1}^{n} W(x_j)y_j\phi_k(x_j)}{\sum_{j=1}^{n} W(x_j)\psi_k^2(x_j)}, \qquad k = 1, 2, \ldots, p. \tag{B5}$$

Another valuable feature of using orthogonal polynomials is seen if we calculate the weighted sum of squared residuals at the minimum. From Eqn (8.12) this is, for a fit of order p,

$$S_p = \frac{1}{\sigma^2}\sum_{j=1}^{n} W(x_j)\left[y_j^2 - \sum_{k=1}^{p} \hat{\theta}_k^2\psi_k^2(x_j)\right], \tag{B6}$$

and thus if we now perform a fit of order $p + 1$, S_p is reduced by

$$\hat{\theta}_{k+1}\sum_{j=1}^{n} W(x_j)\psi_{p+1}^2(k_j),$$

and the first p coefficients $\hat{\theta}_k$ ($k = 1, 2, \ldots, p$) are *unchanged.*

To construct the polynomials we will assume that the values of x_j are normalized to lie in the interval $(-1, 1)$, and since it is also desirable that none of the $\psi_k(x)$ has a large absolute value we will arrange that the leading coefficient of $\psi_k(x)$ is 2^{k-2}. In this case it can be shown that the polynomials satisfy the following recurrence relations. (The derivation of these relations may be found in any modern textbook on numerical analysis).

$$\left.\begin{aligned} &\psi_1(x) = 1/2 \\ &\psi_2(x) = (2x + \beta_1)\psi_1(x) \\ &\text{and for } r \geqslant 2, \\ &\psi_{r+1}(x) = (2x + \beta_r)\psi_r(x) + \gamma_{r-1}\psi_{r-1}(x). \end{aligned}\right\} \tag{B7}$$

To calculate the coefficients β and γ we apply the orthogonality condition to ψ_s and ψ_{r+1}, i.e.

$$\sum_{j=1}^{n} W(x_j)\psi_s(x_j)\psi_{r+1}(x_j) = 0, \qquad s \neq r + 1. \tag{B8}$$

If we use (B7) in (B8) and set $s = j$ and then $j - 1$, we are lead immediately to the results

$$\beta_r = \frac{-2\sum_{j=1}^{n} W(x_j)x_j\psi_r^2(x_j)}{\sum_{j=1}^{n} W(x_j)\psi_r^2(x_j)}, \qquad r = 1, 2, \ldots,$$

and

$$\gamma_{r-1} = \frac{-\sum_{j=1}^{n} W(x_j)\psi_r^2(x_j)}{\sum_{j=1}^{n} W(x_j)\psi_{r-1}^2(x_j)}, \qquad r = 2, 3, \ldots .$$

Appendix C

Optimization of Functions of Several Variables

In Chapters 7 and 8 we encountered the problem of finding the maxima, or minima, of nonlinear functions of several variables. These are examples of a more general class of problems which are currently the subject of intensive research. Strictly, these problems of optimization, although occurring frequently in statistical estimation procedures, are not of a statistical nature, and we confine the discussion of this Appendix to the main ideas involved. For fuller details one should consult one of the excellent books listed in the appropriate section of the bibliography, on which these notes draw extensively.

1 Introduction

We will consider only *minimization* problems since

$$\min f(\mathbf{x}) = \max\,[-f(\mathbf{x})].$$

The general problem to be solved may then be stated as follows. Minimize the *objective function* $f(x_1, x_2, \ldots, x_p) \equiv f(\mathbf{x})$, subject to the m *inequality constraints*

$$c_i(\mathbf{x}) \geqslant 0, \qquad i = 1, 2, \ldots, m$$

and the s *equality constraints,*

$$e_j(\mathbf{x}) = 0, \qquad j = 1, 2, \ldots, s.$$

All other constraints can be reduced to either of the above forms by suitable transformations. We will discuss firstly a few of the features of methods of optimization in general and then, in later sections, discuss in more detail a few of the more successful methods in current use.

Any point which satisfies all the constraints is called *feasible*, and the entire set of such points is called the *feasible region*. Points lying outside the feasible region are said to be *non-feasible*. Nearly all methods of optimization are *iterative* in the sense that an initial feasible vector $\mathbf{x}^{(0)}$ must be specified from which the method will generate a series of vectors $\mathbf{x}^{(1)}$, $\mathbf{x}^{(2)}$, ..., $\mathbf{x}^{(n)}$ etc. which represent improved approximations to the solution. The exception to this occurs in simple tabulation methods where, e.g. in a two-variable problem, a two-dimensional grid of values of $f(x_1, x_2)$ could be calculated and scanned for a minimum. Although systematic methods are available for such multidimensional *grid-searching* far more efficient methods exist for locating optima, and so we shall not discuss them further.

It is convenient to express the iterative procedure by the equation

$$\mathbf{x}^{(n+1)} = \mathbf{x}^{(n)} + h_n \mathbf{d}_n, \tag{C1}$$

where $\mathbf{d}_n$ is a p-dimensional *directional vector*, and h_n is the distance moved along it. The basic problem is to determine the most suitable vector $\mathbf{d}_n$, since once $\mathbf{d}_n$ is chosen the function $f(\mathbf{x})$ can be calculated and a suitable value of h_n found. Iterative techniques fall naturally into two classes, (a) Direct Search Method, and (b) Gradient Methods.

Direct search methods are based on a sequential examination of a series of trial solutions produced from an initial feasible point. On the basis of the examinations, the strategy for further searching is determined. These methods are characterized by the fact they only require explicitly values of the objective function, and, in particular, a knowledge of the derivatives of $f(\mathbf{x})$ is not required. The latter fact is both a strength and a weakness of the methods, for although in problems involving many variables the calculation of derivatives can be difficult, nevertheless it is clear that more efficient methods should be possible if more information (i.e. in the form of the derivatives) is supplied.

In practice direct search methods are most useful for situations involving a few parameters, or where the calculation of derivatives is very difficult, or for finding promising regions in the parameter space where optima might reasonably be located.

Gradient methods, as their name implies, make explicit use of the partial derivatives of the objective function, in addition to values of the function itself. The *gradient direction* at any point is that direction whose components are proportional to the first-order partial derivatives of the objective function at the point. The importance of this quantity will be seen as follows. If we make small perturbations $\delta\mathbf{x}$ from the current point $\mathbf{x}$ then, to first order

$$\delta f = \sum_{j=1}^{p} \frac{\partial f}{\partial x_j} \delta x_j. \tag{C2}$$

To obtain the perturbation giving the greatest change in the function we have to consider the Lagrangian form

$$F(\mathbf{x}, \lambda) = \delta f + \lambda\left(\sum_{j=1}^{p} \delta x_j^2 - \Delta^2\right), \tag{C3}$$

where Δ^2 is the magnitude of the perturbations, i.e.

$$\Delta^2 = \sum_{j=1}^{p} \delta x_j^2.$$

Using (C1) in (C3), and differentiating with respect to δx_j gives

$$\frac{\partial f}{\partial x_j} + 2\lambda\delta x_j = 0,$$

and hence

$$\frac{\delta x_1}{\partial f/\partial x_1} = \frac{\delta x_2}{\partial f/\partial x_2} = \ldots = \frac{\partial x_p}{\partial f/\partial x_p},$$

i.e. for any Δ, the greatest value of δf is obtained if the perturbations δx_j are chosen to be proportional to $\partial f/\partial x_j$, and that, further, if $\delta f < 0$, i.e. the search is to converge to a minimum, the constant of proportionality must be negative. This direction is called the *direction of steepest descent*, and it follows that the objective function can always be reduced by following the direction of steepest descent, although this may only be true for a short distance.

One remark that is worth making about gradient methods concerns the actual calculation of the derivatives. Although gradient methods are, in general, more efficient than direct search methods, their efficiency can drop considerably if the derivatives are not obtained analytically, and so if numerical methods are used to calculate these quantities great care should be exercised to ensure that inaccuracies do not result.

Up to now we have not specified the form of the objective function, except that it is nonlinear in its variables. However, in many practical problems involving unconstrained functions it is found that the objective function can be well-approximated by a quadratic form in the neighbourhood of the minimum. There is therefore considerable interest in methods that *guarantee* to find the minimum of a quadratic in a specified number of steps. Such methods are said to be *quadratically convergent*, and the hope is that problems which are nor strictly quadratic may still be tractable by such methods, a hope which is borne out rather well in practice.

The most useful of the methods having the property of quadratic convergence are those making use of the so-called *conjugate directions* defined as follows.

DEFINITION C.1. Two direction vectors $\mathbf{d}_1$ and $\mathbf{d}_2$ are said to be conjugate with respect to the positive definite matrix $\mathbf{G}$ if

$$\mathbf{d}_1{}^T\mathbf{G}\mathbf{d}_2 = 0.$$

The importance of conjugate directions in optimization problems stems from the following theorem.

THEOREM C.1. *If* $\mathbf{d}_i$ $(i = 1, 2, \ldots, p)$ *are a set of vectors mutually conjugate with respect to the positive definite matrix* $\mathbf{G}$, *then the minimum of the quadratic form*

$$f(\mathbf{x}) = \tfrac{1}{2}\mathbf{x}^T\mathbf{G}\mathbf{x} + \mathbf{b}^T\mathbf{x} + a, \tag{C4}$$

can be found from an arbitrary point $\mathbf{x}^{(0)}$ *by a finite descent calculation in which each of the vectors* $\mathbf{d}_i$ *is used as a descent direction only once, their order of use being arbitrary.*

Proof. Since the set of $\mathbf{d}_i$ are linearly independent any arbitrary vector $\mathbf{v}$ can be written in the form

$$\mathbf{v} = \sum_{i=1}^{p} \alpha_i \mathbf{d}_i, \tag{C5}$$

where, because of the conjugacy of the vectors, $\mathbf{d}_i$,

$$\alpha_i = \frac{\mathbf{d}_i{}^T\mathbf{G}\mathbf{v}}{\mathbf{d}_i{}^T\mathbf{G}\mathbf{d}_i}. \tag{C6}$$

If the general iterative equation (C1) is applied repeatedly, then at the nth stage we have

$$\mathbf{x}^{(n)} = \mathbf{x}^{(0)} + \sum_{i=1}^{n} h_i \mathbf{d}_i, \tag{C7}$$

and in the $(n + 1)$th stage, from $\mathbf{x}^{(n)} \to \mathbf{x}^{(n+1)}$, the distance h_{n+1} along the direction $\mathbf{d}_{n+1}$ is found from the equation

$$\mathbf{d}_{n+1}{}^T\nabla f[\mathbf{x}^{(n+1)}] = 0.$$

Using (C4) this gives

$$\mathbf{d}_{n+1}{}^T\left\{\mathbf{G}\left(\mathbf{x}^{(0)} + \sum_{i=1}^{n} h_i\mathbf{d}_i + h_{n+1}\mathbf{d}_{n+1}\right) + \mathbf{b}\right\} = 0,$$

and so

$$h_{n+1} = \frac{-\mathbf{d}_{n+1}{}^T[\mathbf{Gx}^{(0)} + \mathbf{b}]}{\mathbf{d}_{n+1}{}^T\mathbf{Gd}_{n+1}}, \tag{C8}$$

which depends only on $\mathbf{x}^{(0)}$, and not on the path by which $\mathbf{x}^{(n+1)}$ is reached from $\mathbf{x}^{(0)}$. Using (C8) in (C7) for p-steps gives

$$\mathbf{x}^{(p)} = \mathbf{x}^{(0)} - \sum_{i=1}^{p} \frac{\mathbf{d}_i{}^T[\mathbf{Gx}^{(0)} + \mathbf{b}]\mathbf{d}_i}{\mathbf{d}_i{}^T\mathbf{Gd}_i}, \tag{C9}$$

and using (C5) and (C6) this becomes

$$\mathbf{x}^{(p)} = \mathbf{x}^{(0)} - \mathbf{x}^{(0)} - \mathbf{G}^{-1}\mathbf{b} = -\mathbf{G}^{-1}\mathbf{b}, \tag{C10}$$

which shows that the minimum has been reached in p iterations.

Although methods having the property of quadratic convergence will guarantee to converge to the exact minimum of a quadratic in p steps, where p is the dimensionality of the problem, when applied to functions which are not strictly quadratic the problem arises of determining when convergence has taken place. This is, in general, a difficult problem, but in practice a suitable criterion is to consider that convergence has been achieved if, for given small values of ε and ε'

$$f(\mathbf{x}^{(n)}) - f(\mathbf{x}^{(n+1)}) < \varepsilon,$$

and/or

$$|\mathbf{x}^{(n)} - \mathbf{x}^{(n+1)}| < \varepsilon',$$

for a sequence of q successive interations, where q is a number which will vary with the type of function being minimized, but at a very generous over estimate $q \sim p$, the number of variables. In practice a considerably smaller number of values is usually sufficient.

Finally, it should be mentioned that all present techniques for optimizing nonlinear functions locate only *local optima*, i.e. points $\mathbf{x}_m$ at which $f(\mathbf{x}_m) < f(\mathbf{x})$ for all $\mathbf{x}$ in a region in the neighbourhood of $\mathbf{x}_m$. For multivariate problems there could well be better local optima located at some distance from $\mathbf{x}_m$. At present there are no general methods for locating the *global optimum* (i.e. the absolute optimum) of a function, and so it is essential to restart the search procedure from many different initial points $\mathbf{x}^{(0)}$ to ensure that the full p-dimensional space has been explored.

2 Unconstrained minimization

We will start the discussion of specific techniques by considering the unconstrained problem, for which powerful general techniques have been devised.

C.2.1. Univariate Problems

The problem of minimizing a function of one variable is very important in practice, because, as we shall see later, many methods for optimizing multivariate functions proceed by a series of searches along a line in the parameter space, and each of these searches is equivalent to a univariate search.

Univariate searches fall into two groups (a) those which specify an interval within which the minimum lies, and (b) those which specify the minimum by a point approximating it. The latter methods are the most useful in practice and we shall only consider them here. The basic procedure is as follows. Proceeding from an initial point $x^{(0)}$ a systematic search technique is applied to find a region containing the minimum. This bracket is then refined by fitting a quadratic interpolation polynomial to the three points making up the bracket, and locating the minimum of this polynomial. As a result of this evaluation a new bracket is formed, and the procedure is repeated. This method is both simple and very safe in practice.

As an example of the above technique we will give the method of Davies, Swann and Campey.

(a) *Method of Davies, Swann and Campey*

To bracket the minimum the function is first evaluated at $f(x^{(0)})$ and $f(x^{(0)}+h)$. If $f(x^{(0)}+h) \leqslant f(x^{(0)})$, then $f(x^{(0)}+2h)$ is evaluated. This doubling of the step-length h is repeated until a value of $f(x)$ is found such that $f(x^{(0)}+2^n h) > f(x^{(0)}+2^{n-1}h)$. At this point the step-length is halved and a step again taken from the last successive point, i.e. the $(n-1)$th. This procedure produces four points equally spaced along the axis of search, at each of which the function has been evaluated. The end point furthest from the point corresponding to the smallest function value is rejected, and the remaining three points used for quadratic interpolation. Had the first step failed then the search is continued by reversing the sign of the step length. If the first step in this direction also fails then the minimum has been bracketed and the interpolation may be made. If the three points used for the interpolation are x_1, x_2, x_3 with $x_1 < x_2 < x_3$ and $x_3 - x_2 = x_2 - x_1 = l$, then the minimum of the fitted quadratic is at

$$x_m = x_2 + \frac{l[f(x_1) - f(x_3)]}{2[f(x_1) - 2f(x_2) + f(x_3)]}.$$

An iteration is completed by evaluating $f(x_m)$. Convergence tests are now applied and, if required, a further iteration is performed, with a reduced step length, using as the initial point whichever of x_2 or x_m corresponds to the smaller function value.

3 Multivariate problems

C.3.1. Direct Search Methods

As we mentioned in Section C.2.1. univariate searches are important as many methods for locating optima of multivariate functions are based on a series of linear searches along a line in the parameter space. By a *linear method* we will therefore mean any technique which uses a set of direction vectors in the search, and which proceeds by explorations along these directions, deciding future strategy by the results obtained in previous searches.

The simplest of all possible methods would be to keep $(p - 1)$ of the parameters fixed and find a minimum with respect to the pth parameter, doing this in turn for each variable. The progress of such an *alternating-variable* search is shown in Fig. C.1 for the case of two variables. In general the contours of equal function value will be aligned along the so-called *principal axes*, which are not parallel to the coordinate axes. In this case only very small steps will be taken at each stage and the technique is very inefficient, being worse the larger the number of variables. It would clearly be very much more efficient to reorientate the direction vectors along the principal axes, and this is done in several techniques, one of the most successful of which is due to Rosenbrock.

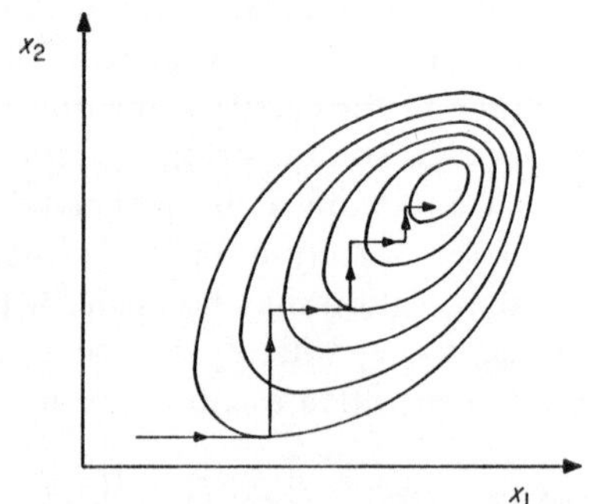

Fig. C.1. Typical progress in an alternating variable search.

(a) *Rosenbrock's Method*

The method of Rosenbrock utilizes p orthonormal direction vectors, which are initially taken to be the coordinate directions, and reorientates them during the search procedure such that one of them lies in the direction of total recent progress of the search. We will denote the p direction vectors at the nth stage by $\mathbf{d}_i^{(n)}(i = 1, 2, \ldots, p)$, and by h_i the step length associated with each of these directions. A single stage of the method is then as follows.

A step h_i is taken in the direction $\mathbf{d}_i^{(n)}$ from a given starting point $\mathbf{x}^{(n-1)}$. If $f(\mathbf{x}^{(n)}) \leqslant f(\mathbf{x}^{(n-1)})$ the step is considered successful and h_i is multiplied by a fixed multiplier $\alpha > 1$, the new point being kept and the success recorded. If $f(\mathbf{x}^{(n)}) > f(\mathbf{x}^{(n-1)})$ the step is retracted, h_i is multiplied by β, where $-1 < \beta < 0$, and a failure is recorded. This procedure is repeated for all variables in turn from 1 to p and then starting again at 1, cycling until a success followed by a failure is recorded along every one of the p directions. A new set of orthonormal vectors is now constructed as follows.

If s_i is the algebraic sum of all the successful steps in the direction $\mathbf{d}_i^{(n}$ during the nth stage, we can define vectors $\mathbf{a}_i$ $(i = 1, 2, \ldots, p)$ by

$$\mathbf{a}_i = \sum_{k=i}^{p} s_k \mathbf{d}_k^{(n)},$$

so that $\mathbf{a}_1$ represents the total progress made during the complete stage, $\mathbf{a}_2$ the total progress made *excluding* that made in the direction $\mathbf{d}_1^{(n)}$ etc. Since the set $\mathbf{a}_i$ are linearly independent they can be used to generate a new set of orthonormal directions by the Gram–Schmidt procedure which gives

$$\mathbf{d}_k' = \mathbf{a}_k - \sum_{l=1}^{k-1} (\mathbf{a}_k^T \mathbf{d}_l^{(n+1)}) \mathbf{d}_l^{(n+1)},$$

with

$$\mathbf{d}_k^{(n+1)} = \mathbf{d}_k' / |\mathbf{d}_k'|,$$

and $k = 1, 2, \ldots, p$. Thus a new stage can be started with the new set of direction vectors $\mathbf{d}_i^{(n+1)}$, and the procedure is repeated until some suitable convergence criteria are satisfied.

(b) *Powell's Method*

A more powerful method which is based on the use of conjugate directions rather than the orthonormal vectors of Rosenbrock's method is due to Powell. It uses the fact that, for a positive definite quadratic form, if searches for minima are made along p conjugate directions then the join of these minima is conjugate to all of those directions, a result that follows from Theorem C.1.

In the usual notation, the procedure is to start from $\mathbf{x}^{(0)}$ and locate the minimum $\mathbf{x}^{(1)}$ in the direction $\mathbf{d}_1^{(n)}$. Then from $\mathbf{x}^{(1)}$ locate the minimum $\mathbf{x}^{(2)}$ in the direction $\mathbf{d}_2^{(n)}$ etc. until the minimum $\mathbf{x}^{(p)}$ in the direction $\mathbf{d}_p^{(n)}$ is found. The direction of total progress made during this cycle is then

$$\mathbf{d} = \mathbf{x}^{(p)} - \mathbf{x}^{(0)}.$$

Provided that certain conditions hold the minimum in the direction $\mathbf{d}$ is then found and used as a starting point in the next iteration, the list of direction vectors being updated as follows

$$\begin{aligned}&(\mathbf{d}_1^{(n+1)}, \mathbf{d}_2^{(n+1)}, \ldots, \mathbf{d}_p^{(n+1)})\\ &= (\mathbf{d}_1^{(n)}, \mathbf{d}_2^{(n)}, \ldots, \mathbf{d}_{j-1}^{(n)}, \mathbf{d}_{j+1}^{(n)}, \ldots, \mathbf{d}_p^{(n)}, \mathbf{d})\end{aligned}$$

where $\mathbf{d}_j^{(n)}$ is that direction vector along which the greatest reduction in the function value occured during the nth stage. Care must be taken in this updating procedure to ensure that the new direction vectors are always linearly independent.

Powell shows that for the quadratic function of Eqn (C4), if $\mathbf{d}_i^{(n)}$ is scaled so that

$$\mathbf{d}_i^{(n)T}\mathbf{G}\mathbf{d}_i^{(n)} = 1, \qquad i = 1, 2, \ldots, p,$$

then the determinant D of the matrix whose columns are $\mathbf{d}_i^{(n)}$ has a maximum if, and only if, the vectors $\mathbf{d}_i^{(n)}$ are mutually conjugate with respect to $\mathbf{G}$. Thus the direct $\mathbf{d}$ only replaces an existing search direction if by so doing D is increased.

C.3.2. Gradient Methods

The earliest method using gradients is that of steepest descent mentioned previously. In this method the normalized gradient vector at the current point is found, and using a step-length h_i a new point is generated via the general iterative equation. This procedure is continued until a function value is found which has not decreased. The step-length is then reduced and the search restarted from the best previous point. If the actual minimum along each search direction is located then the performance of this method is similar in appearance to the alternating variable search, and, in particular, is rather erratic, the search directions oscillating about the principal axes. A method that in principle is far better is based on an examination of the second derivatives of the function.

(a) *Newton's Method*

A second-order Taylor expansion of the objective function $f(x)$ about the minimum $\mathbf{x}^*$ is

$$f(\mathbf{x}) = f(\mathbf{x}^*) + \sum_{j=1}^{p} h_j \left(\frac{\partial f}{\partial x_j}\right)^* + \tfrac{1}{2} \sum_{j=1}^{p} \sum_{k=1}^{p} h_j h_k \left(\frac{\partial^2 f}{\partial x_j \partial x_k}\right)^*,$$

where * means that these quantities are evaluated at $\mathbf{x} = \mathbf{x}^*$. Differentiating this equation gives

$$g_l \equiv \frac{\partial f}{\partial x_l} = \sum_{j=1}^{p} h_j \left(\frac{\partial^2 f}{\partial x_j \partial x_l}\right)^*, \qquad l = 1, 2, \ldots, p. \tag{C11}$$

The minimum is therefore obtained in one step by the move $\mathbf{x}^* = \mathbf{x} - \mathbf{h}$, where the components of $\mathbf{h}$ are found by solving the p linear equations (C11). If we define

$$G_{jk} \equiv \frac{\partial^2 f}{\partial x_j \partial x_k},$$

then we have

$$\mathbf{x}^* = \mathbf{x} - \mathbf{G}^{*-1}\mathbf{g}.$$

Since $\mathbf{G}^{*-1}$ will not, of course, be known it is usual to replace it by $\mathbf{G}^{-1}$ evaluated at the current point $\mathbf{x}^{(i)}$ and use the iterative equation

$$\mathbf{x}^{(n+1)} = \mathbf{x}^{(n)} - \mathbf{G}_n^{-1}\mathbf{g}_n. \tag{C12}$$

The method is clearly quadratically convergent but suffers from severe difficulties.

Firstly, there is the numerical problem of calculating the inverse matrix of second derivatives, and secondly, and more seriously, for a general function $\mathbf{G}^{-1}$ is not guaranteed to be positive definite, and in this case the method will diverge. Thus, while Newton's method is efficient in the immediate neighbourhood of a minimum, away from this point it has little to recommend it, the method of steepest descent being far preferable.

In view of the above remarks an efficient method would be one that starts by using the method of steepest descent and, at a later stage, uses Newton's method. A method which does this automatically is due to Davidon, and probably represents the most powerful method currently available for optimizing unconstrained functions.

(b) *Davidon's Method*

Davidon's method is an iterative scheme based on successive approximations to the matrix $\mathbf{G}^{*-1}$. The best approximation to this matrix, say $\mathbf{H}_n$, is used to define a new search direction by a modification of Eqn (C12), i.e.

$$\mathbf{x}^{(n+1)} = \mathbf{x}^{(n)} - h_n \mathbf{H}_n \mathbf{g}_n,$$

where $\mathbf{g}_n$ is the vector of first derivatives of $f(\mathbf{x}^{(n)})$ with respect to $\mathbf{x}^{(n)}$. The step length h_n is that necessary to find the minimum in the search direction $\mathbf{d}_n = -\mathbf{H}_n \mathbf{g}_n$, and may be found by any univariate search procedure. If the sequence $\{\mathbf{H}_n\}$ is positive definite, it can be shown that the convergence of this method is guaranteed. Furthermore, if the search directions $\mathbf{d}_n$ are mutually conjugate, then the method is quadratically convergent.

Davidon has shown that both of these conditions can be met if, at each stage of the iteration, the matrix $\mathbf{H}_n$ is updated according to the equation

$$\mathbf{H}_{n+1} = \mathbf{H}_n + \mathbf{A}_n + \mathbf{B}_n,$$

where

$$\mathbf{A}_n = \frac{-h_n[\mathbf{H}_n \mathbf{g}_n \mathbf{g}_n{}^T \mathbf{H}_n{}^T]}{(\mathbf{H}_n \mathbf{g}_n)^T \mathbf{V}},$$

$$\mathbf{B}_n = \frac{-\mathbf{H}_n \mathbf{V} \mathbf{V}^T \mathbf{H}_n}{\mathbf{V}^T \mathbf{H}_n \mathbf{V}},$$

and

$$\mathbf{V} = \mathbf{g}_{n+1} - \mathbf{g}_n.$$

It is usual to start the iteration from $\mathbf{H}_0 = \mathbf{1}$. The matrix $\mathbf{A}_n$ ensures that the sequence $\{\mathbf{H}_n\}$ converges to $\mathbf{G}^{*-1}$, and $\mathbf{B}_n$ ensures that each $\mathbf{H}_n$ is positive definite. The derivation of these expressions may be found in the book by Kowalik and Osborne, cited in the Bibliography.

4 Constrained minimization

Constrained optimization, not surprisingly, is a more difficult problem than unconstrained optimization, and only a very brief discussion will be given here.

Firstly, an obvious remark. If the constraints can be removed by suitable transformations then this, of course, should be done. For example many problems involve the simple parameter constraints

$$l \leqslant x \leqslant u,$$

which can be removed entirely by the transformation

$$x = l + (u - l)\sin^2 y,$$

thereby enabling an unconstrained minimization to be performed with respect to y. Such transformations *cannot* produce additional local optima.

If the constraints cannot be removed then one of the simplest ways of incorporating them is to arrange that the production of non-feasible points is unattractive. This is the basis of the method of *penalty functions*.

C.4.1. Penalty Functions

If the problem is to minimize $f(\mathbf{x})$ subject to the m inequality constraints

$$c_i(\mathbf{x}) \geqslant 0, \qquad i = 1, 2, \ldots, m,$$

and the s equality constraints

$$e_j(\mathbf{x}) = 0, \qquad j = 1, 2, \ldots, s,$$

we could consider the function

$$F(\mathbf{x}) = f(\mathbf{x}) + \sum_{i=1}^{m} \lambda_i c_i^2(\mathbf{x}) H[c_i(\mathbf{x})] + \sum_{j=1}^{s} \lambda_j' e_j^2(\mathbf{x}), \tag{C13}$$

where $H(q)$ is the (0, 1) step function

$$H(q) = \begin{cases} 0, & q \geqslant 0 \\ 1, & q < 0, \end{cases}$$

and λ_i, λ_i' are positive scaling factors, chosen so that the contributions of the various terms to (C13) are approximately equal. The penalty is thus the weighted sum of squares of the amount by which the constraints are violated.

This method works reasonably well in practice, but has the disadvantage of requiring that values of $f(\mathbf{x})$ be calculated at non-feasible points, and this may not always be possible. A method which restricts the search to feasible points only is due to Carroll, and is known as *Carroll's created response surface technique*. In this method if the constraints are inequalities, the surface

$$F(\mathbf{x}, k) = f(\mathbf{x}) + k \sum_{i=1}^{m} \frac{1}{c_i(\mathbf{x})},$$

is considered, where $k > 0$, and a minimum found as a function of $\mathbf{x}$. This minimum point is then used as the starting value for a new minimization for a reduced value of k, and the procedure repeated until $k = 0$ is reached. In all minimizations non-feasible points are excluded. The theoretical development of this method, and its extension to incorporate equality constraints may be found in the book of Kowalik and Osborne.

Appendix D

Statistical Tables

TABLE D1

Ordinates of the standard normal density function

$$f(x) = \frac{1}{(2\pi)^{\frac{1}{2}}} \exp(-x^2/2).$$

Note that

$$f(-x) = f(x).$$

x	·00	·01	·02	·03	·04	·05	·06	·07	·08	·09
·0	·3989	·3989	·3989	·3988	·3986	·3984	·3982	·3980	·3977	·3973
·1	·3970	·3965	·3961	·3956	·3951	·3945	·3939	·3932	·3925	·3918
·2	·3910	·3902	·3894	·3885	·3876	·3867	·3857	·3847	·3836	·3825
·3	·3814	·3802	·3790	·3778	·3765	·3752	·3739	·3725	·3712	·3697
·4	·3683	·3668	·3653	·3637	·3621	·3605	·3589	·3572	·3555	·3538
·5	·3521	·3503	·3485	·3467	·3448	·3429	·3410	·3391	·3372	·3352
·6	·3332	·3312	·3292	·3271	·3251	·3230	·3209	·3187	·3166	·3144
·7	·3123	·3101	·3079	·3056	·3034	·3011	·2989	·2966	·2943	·2920
·8	·2897	·2874	·2850	·2827	·2803	·2780	·2756	·2732	·2709	·2685
·9	·2661	·2637	·2613	·2589	·2565	·2541	·2516	·2492	·2468	·2444
1·0	·2420	·2396	·2371	·2347	·2323	·2299	·2275	·2251	.2227	·2205
1·1	·2179	·2155	·2131	·2107	·2083	·2059	·2036	·2012	·1989	·1963
1·2	·1942	·1919	·1895	·1872	·1849	·1826	·1804	·1781	·1758	·1735
1·3	·1714	·1691	·1669	·1647	·1626	·1604	·1582	·1561	·1539	·1516
1·4	·1497	·1476	·1456	·1435	·1415	·1394	·1374	·1354	·1334	·1318
1·5	·1295	·1276	·1257	·1238	·1219	·1200	·1182	·1163	·1145	·1127
1·6	·1109	·1092	·1074	·1057	·1040	·1023	·1006	·0989	·0973	·0957
1·7	·0940	·0925	·0909	·0893	·0878	·0863	·0848	·0833	·0818	·0804
1·8	·0790	·0775	·0761	·0748	·0734	·0721	·0707	·0694	·0681	·0669
1·9	·0656	·0644	·0632	·0620	·0608	·0596	·0584	·0573	·0562	·0551
2·0	·0540	·0529	·0519	·0508	·0498	·0488	·0478	·0468	·0459	·0449
2·1	·0440	·0431	·0422	·0413	·0404	·0396	·0387	·0379	·0371	·0363

TABLE D1 (*continued*)

x	·00	·01	·02	·03	·04	·05	·06	·07	·08	·09
2·2	·0355	·0347	·0339	·0332	·0325	·0317	·0310	·0303	·0297	·0290
2·3	·0283	·0277	·0270	·0264	·0258	·0252	·0246	·0241	·0235	·0229
2·4	·0224	·0219	·0213	·0208	·0203	·0198	·0194	·0189	·0184	·0180
2·5	·0175	·0171	·0167	·0163	·0158	·0154	·0151	·0147	·0143	·0139
2·6	·0136	·0132	·0129	·0126	·0122	·0119	·0116	·0113	·0110	·0107
2·7	·0104	·0101	·0099	·0096	·0093	·0091	·0088	·0086	·0084	·0081
2·8	·0079	·0077	·0075	·0073	·0071	·0069	·0067	·0065	·0063	·0061
2·9	·0060	·0058	·0056	·0055	·0053	·0051	·0050	·0048	·0047	·0046
3·0	·0044	·0043	·0042	·0040	·0039	·0038	·0037	·0036	·0035	·0034
3·1	·0033	·0032	·0031	·0030	·0029	·0028	·0027	·0026	·0025	·0025
3·2	·0024	·0023	·0022	·0022	·0021	·0020	·0020	·0019	·0018	·0018
3·3	·0017	·0017	·0016	·0016	·0015	·0015	·0014	·0014	·0013	·0013
3·4	·0012	·0012	·0012	·0011	·0011	·0010	·0010	·0010	·0009	·0009
3·5	·0009	·0008	·0008	·0008	·0008	·0007	·0007	·0007	·0007	·0006
3·6	·0006	·0006	·0006	·0005	·0005	·0005	·0005	·0005	·0005	·0004
3·7	·0004	·0004	·0004	·0004	·0004	·0004	·0003	·0003	·0003	·0003
3·8	·0003	·0003	·0003	·0003	·0003	·0002	·0002	·0002	·0002	·0002
3·9	·0002	·0002	·0002	·0002	·0002	·0002	·0002	·0002	·0001	·0001

TABLE D2

Standard normal distribution function

$$F(x) = \frac{1}{(2\pi)^{\frac{1}{2}}} \int_{-\infty}^{x} dt \exp(-t^2/2).$$

Note that

$$F(-x) = 1 - F(x).$$

x	·00	·01	·02	·03	·04	·05	·06	·07	·08	·09
·0	·5000	·5040	·5080	·5120	·5160	·5199	·5239	·5279	·5319	·5359
·1	·5398	·5438	·5478	·5517	·5557	·5596	·5636	·5675	·5714	·5753
·2	·5793	·5832	·5871	·5910	·5948	·5987	·6026	·6064	·6103	·6141
·3	·6179	·6217	·6255	·6293	·6331	·6368	·6406	·6443	·6480	·6517
·4	·6554	·6591	·6628	·6664	·6700	·6736	·6772	·6808	·6844	·6879
·5	·6915	·6950	·6985	·7019	·7054	·7088	·7123	·7157	·7190	·7224
·6	·7257	·7291	·7324	·7357	·7389	·7422	·7454	·7486	·7517	·7349
·7	·7580	·7611	·7642	·7673	·7704	·7734	·7764	·7794	·7823	·7852
·8	·7881	·7910	·7939	·7967	·7995	·8023	·8051	·8078	·8106	·8133
·9	·8159	·8186	·8212	·8238	·8264	·8289	·8315	·8340	·8365	·8389
1·0	·8413	·8438	·8461	·8485	·8508	·8531	·8554	·8577	·8599	·8621
1·1	·8643	·8665	·8686	·8708	·8729	·8749	·8770	·8790	·8810	·8830
1·2	·8849	·8869	·8888	·8907	·8925	·8944	·8962	·8980	·8997	·9015
1·3	·9032	·9049	·9066	·9082	·9099	·9115	·9131	·9147	·9162	·9177
1·4	·9192	·9207	·9222	·9236	·9251	·9265	·9279	·9292	·9306	·9319
1·5	·9332	·9345	·9357	·9370	·9382	·9394	·9406	·9418	·9429	·9441
1·6	·9452	·9463	·9474	·9484	·9495	·9505	·9515	·9525	·9535	·9545
1·7	·9554	·9564	·9573	·9582	·9591	·9599	·9608	·9616	·9625	·9633
1·8	·9641	·9649	·9656	·9664	·9671	·9678	·9686	·9693	·9699	·9706
1·9	·9713	·9719	·9726	·9732	·9738	·9744	·9750	·9756	·9761	·9767
2·0	·9772	·9778	·9783	·9788	·9793	·9798	·9803	·9808	·9812	·9817
2·1	·9821	·9826	·9830	·9834	·9838	·9842	·9846	·9850	·9854	·9857
2·2	·9861	·9864	·9868	·9871	·9875	·9878	·9881	·9884	·9887	·9890
2·3	·9893	·9896	·9898	·9901	·9904	·9906	·9909	·9911	·9913	·9916
2·4	·9918	·9920	·9922	·9925	·9927	·9929	·9931	·9932	·9934	·9936
2·5	·9938	·9940	·9941	·9943	·9945	·9946	·9948	·9949	·9951	·9952
2·6	·9953	·9955	·9956	·9957	·9959	·9960	·9961	·9962	·9963	·9964
2·7	·9965	·9966	·9967	·9968	·9969	·9970	·9971	·9972	·9973	·9974
2·8	·9974	·9975	·9976	·9977	·9977	·9978	·9979	·9979	·9980	·9981
2·9	·9981	·9982	·9982	·9983	·9984	·9984	·9985	·9985	·9986	·9986
3·0	·9987	·9987	·9987	·9988	·9988	·9989	·9989	·9989	·9990	·9990
3·1	·9990	·9991	·9991	·9991	·9992	·9992	·9992	·9992	·9993	·9993
3·2	·9993	·9993	·9994	·9994	·9994	·9994	·9994	·9995	·9995	·9995
3·3	·9995	·9995	·9995	·9996	·9996	·9996	·9996	·9996	·9996	·9997
3·4	·9997	·9997	·9997	·9997	·9997	·9997	·9997	·9997	·9997	·9998

x	1·282	1·645	1·960	2·326	2·576	3·090	3·291	3·891	4·417
$F(x)$	·90	·95	·975	·99	·995	·999	·9995	·99995	·999995
$2[1-F(x)]$	·20	·10	·05	·02	·01	·002	·001	·0001	·00001

TABLE D3

The probability of exactly r successes in n independent Bernoulli trials, with the probability of a success in a single trial equal to p, is given by the $(r+1)$th term in the binomial expansion of $(q+p)^n$, i.e.

$$f(x) = \binom{n}{r} p^r q^{n-r}, \qquad r = 0, 1, \ldots, n \qquad (q+p) = 1.$$

The probability of obtaining s or more successes is given by

$$F = \sum_{r=s}^{n} \binom{n}{r} p^r q^{n-r}.$$

This table gives values of F for specified values of n, s and p. If $p > 0{\cdot}5$ the values for F are obtained from

$$1 - \sum_{r=n-s+1}^{n} \binom{n}{r} q^r p^{n-r}.$$

		p									
n	*s*	·05	·10	·15	·20	·25	·30	·35	·40	·45	·50
2	1	·0975	·1900	·2775	·3600	·4375	·5100	·5775	·6400	·6975	·7500
	2	·0025	·0100	·0225	·0400	·0625	·0900	·1225	·1600	·2025	·2500
3	1	·1426	·2710	·3859	·4880	·5781	·6570	·7254	·7840	·8336	·8750
	2	·0072	·0280	·0608	·1040	·1562	·2160	·2818	·3520	·4252	·5000
	3	·0001	·0010	·0034	·0080	·0156	·0270	·0429	·0640	·0911	·1250
4	1	·1855	·3439	·4780	·5904	·6836	·7599	·8215	·8704	·9085	·9375
	2	·0140	·0523	·1095	·1808	·2617	·3483	·4370	·5248	·6090	·6875
	3	·0005	·0037	·0120	·0272	·0508	·0837	·1265	·1792	·2415	·3125
	4	·0000	·0001	·0005	·0016	·0039	·0081	·0150	·0256	·0410	·0625
5	1	·2262	·4095	·5563	·6723	·7627	·8319	·8840	·9222	·9497	·9688
	2	·0226	·0815	·1648	·2627	·3672	·4718	·5716	·6630	·7438	·8125
	3	·0012	·0086	·0266	·0579	·1035	·1631	·2352	·3174	·4069	·5000
	4	·0000	·0005	·0022	·0067	·0156	·0308	·0540	·0870	·1312	·1875
	5	·0000	·0000	·0001	·0003	·0010	·0024	·0053	·0102	·0185	·0312
6	1	·2649	·4686	·6229	·7379	·8220	·8824	·9246	·9533	·9723	·9844
	2	·0328	·1143	·2235	·3447	·4661	·5798	·6809	·7667	·8364	·8906
	3	·0022	·0158	·0473	·0989	·1694	·2557	·3529	·4557	·5585	·6562
	4	·0001	·0013	·0059	·0170	·0376	·0705	·1174	·1792	·2553	·3438
	5	·0000	·0001	·0004	·0016	·0046	·0109	·0223	·0410	·0692	·1094
	6	·0000	·0000	·0000	·0001	·0002	·0007	·0018	·0041	·0083	·0156
7	1	·3017	·5217	·6794	·7903	·8665	·9176	·9510	·9720	·9848	·9922
	2	·0444	·1497	·2834	·4233	·5551	·6706	·7662	·8414	·8976	·9375

TABLE D3 (*continued*)

n	*s*	*p* ·05	·10	·15	·20	·25	·30	·35	·40	·45	·50
	3	·0038	·0257	·0738	·1480	·2436	·3529	·4677	·5801	·6836	·7734
	4	·0002	·0027	·0121	·0333	·0706	·1260	·1998	·2898	·3917	·5000
	5	·0000	·0002	·0012	·0047	·0129	·0288	·0556	·0963	·1529	·2266
	6	·0000	·0000	·0001	·0004	·0013	·0038	·0090	·0188	·0357	·0625
	7	·0000	·0000	·0000	·0000	·0001	·0002	·0006	·0016	·0037	·0078
8	1	·3366	·5695	·7275	·8322	·8999	·9424	·9681	·9832	·9916	·9961
	2	·0572	·1869	·3428	·4967	·6329	·7447	·8309	·8936	·9368	·9648
	3	·0058	·0381	·1052	·2031	·3215	·4482	·5722	·6846	·7799	·8555
	4	·0004	·0050	·0214	·0563	·1138	·1941	·2936	·4059	·5230	·6367
	5	·0000	·0004	·0029	·0104	·0273	·0580	·1061	·1737	·2604	·3633
	6	·0000	·0000	·0002	·0012	·0042	·0113	·0253	·0498	·0885	·1445
	7	·0000	·0000	·0000	·0001	·0004	·0013	·0036	·0085	·0181	·0352
	8	·0000	·0000	·0000	·0000	·0000	·0001	·0002	·0007	·0017	·0039
9	1	·3698	·6126	·7684	·8658	·9249	·9596	·9793	·9899	·9954	·9980
	2	·0712	·2252	·4005	·5638	·6997	·8040	·8789	·9295	·9615	·9805
	3	·0084	·0530	·1409	·2618	·3993	·5372	·6627	·7682	·8505	·9102
	4	·0006	·0083	·0339	·0856	·1657	·2703	·3911	·5174	·6386	·7461
	5	·0000	·0009	·0056	·0196	·0489	·0988	·1717	·2666	·3786	·5000
	6	·0000	·0001	·0006	·0031	·0100	·0253	·0536	·0994	·1658	·2539
	7	·0000	·0000	·0000	·0003	·0013	·0043	·0112	·0250	·0498	·0898
	8	·0000	·0000	·0000	·0000	·0001	·0004	·0014	·0038	·0091	·0195
	9	·0000	·0000	·0000	·0000	·0000	·0000	·0001	·0003	·0008	·0020
10	1	·4013	·6513	·8031	·8926	·9437	·9718	·9865	·9940	·9975	·9990
	2	·0861	·2639	·4557	·6242	·7560	·8507	·9140	·9536	·9767	·9893
	3	·0115	·0702	·1798	·3222	·4744	·6172	·7384	·8327	·9004	·9453
	4	·0010	·0128	·0500	·1209	·2241	·3504	·4862	·6177	·7340	·8281
	5	·0001	·0016	·0099	·0328	·0781	·1503	·2485	·3669	·4956	·6230
	6	·0000	·0001	·0014	·0064	·0197	·0473	·0949	·1662	·2616	·3770
	7	·0000	·0000	·0001	·0009	·0035	·0106	·0260	·0548	·1020	·1719
	8	·0000	·0000	·0000	·0001	·0004	·0016	·0048	·0123	·0274	·0547
	9	·0000	·0000	·0000	·0000	·0000	·0001	·0005	·0017	·0045	·0107
	10	·0000	·0000	·0000	·0000	·0000	·0000	·0000	·0001	·0003	·0010
11	1	·4312	·6862	·8327	·9141	·9578	·9802	·9912	·9964	·9986	·9995
	2	·1019	·3026	·5078	·6779	·8029	·8870	·9394	·9698	·9861	·9941
	3	·0152	·0896	·2212	·3826	·5448	·6873	·7999	·8811	·9348	·9673
	4	·0016	·0185	·0694	·1611	·2867	·4304	·5744	·7037	·8089	·8867
	5	·0001	·0028	·0159	·0504	·1146	·2103	·3317	·4672	·6029	·7256
	6	·0000	·0003	·0027	·0117	·0343	·0782	·1487	·2465	·3669	·5000
	7	·0000	·0000	·0003	·0020	·0076	·0216	·0501	·0994	·1738	·2744
	8	·0000	·0000	·0000	·0002	·0012	·0043	·0122	·0293	·0610	·1133
	9	·0000	·0000	·0000	·0000	·0001	·0006	·0020	·0059	·0148	·0327

TABLE D3 (*continued*)

n	s	·05	·10	·15	·20	·25	·30	·35	·40	·45	·50
	10	·0000	·0000	·0000	·0000	·0000	·0000	·0002	·0007	·0022	·0059
	11	·0000	·0000	·0000	·0000	·0000	·0000	·0000	·0000	·0002	·0005
12	1	·4596	·7176	·8578	·9313	·9683	·9862	·9943	·9978	·9992	·9998
	2	·1184	·3410	·5565	·7251	·8416	·9150	·9576	·9804	·9917	·9968
	3	·0196	·1109	·2642	·4417	·6093	·7472	·8487	·9166	·9579	·9807
	4	·0022	·0256	·0922	·2054	·3512	·5075	·6533	·7747	·8655	·9270
	5	·0002	·0043	·0239	·0726	·1576	·2763	·4167	·5618	·6956	·8062
	6	·0000	·0005	·0046	·0194	·0544	·1178	·2127	·3348	·4731	·6128
	7	·0000	·0001	·0007	·0039	·0143	·0386	·0846	·1582	·2607	·3872
	8	·0000	·0000	·0001	·0006	·0028	·0095	·0255	·0573	·1117	·1938
	9	·0000	·0000	·0000	·0001	·0004	·0017	·0056	·0153	·0356	·0730
	10	·0000	·0000	·0000	·0000	·0000	·0002	·0008	·0028	·0079	·0193
	11	·0000	·0000	·0000	·0000	·0000	·0000	·0001	·0003	·0011	·0032
	12	·0000	·0000	·0000	·0000	·0000	·0000	·0000	·0000	·0001	·0002
13	1	·4867	·7458	·8791	·9450	·9762	·9903	·9963	·9987	·9996	·9999
	2	·1354	·3787	·6017	·7664	·8733	·9363	·9704	·9874	·9951	·9983
	3	·0245	·1339	·2704	·4983	·6674	·7975	·8868	·9421	·9731	·9888
	4	·0031	·0342	·0967	·2527	·4157	·5794	·7217	·8314	·9071	·9539
	5	·0003	·0065	·0260	·0991	·2060	·3457	·4995	·6470	·7721	·8666
	6	·0000	·0009	·0053	·0300	·0802	·1654	·2841	·4256	·5732	·7095
	7	·0000	·0001	·0013	·0070	·0243	·0624	·1295	·2288	·3563	·5000
	8	·0000	·0000	·0002	·0012	·0056	·0182	·0462	·0977	·1788	·2905
	9	·0000	·0000	·0000	·0002	·0010	·0040	·0126	·0321	·0698	·1334
	10	·0000	·0000	·0000	·0000	·0001	·0007	·0025	·0078	·0203	·0461
	11	·0000	·0000	·0000	·0000	·0000	·0001	·0003	·0013	·0041	·0112
	12	·0000	·0000	·0000	·0000	·0000	·0000	·0000	·0001	·0005	·0017
	13	·0000	·0000	·0000	·0000	·0000	·0000	·0000	·0000	·0000	·0001
14	1	·5123	·7712	·8972	·9560	·9822	·9932	·9976	·9992	·9998	·9999
	2	·1530	·4154	·6433	·8021	·8990	·9525	·9795	·9919	·9971	·9991
	3	·0301	·1584	·3521	·5519	·7189	·8392	·9161	·9602	·9830	·9935
	4	·0042	·0441	·1465	·3018	·4787	·6448	·7795	·8757	·9368	·9713
	5	·0004	·0092	·0467	·1298	·2585	·4158	·5773	·7207	·8328	·9102
	6	·0000	·0015	·0115	·0439	·1117	·2195	·3595	·5141	·6627	·7880
	7	·0000	·0002	·0022	·0116	·0383	·0933	·1836	·3075	·4539	·6047
	8	·0000	·0000	·0003	·0024	·0103	·0315	·0753	·1501	·2586	·3953
	9	·0000	·0000	·0000	·0004	·0022	·0083	·0243	·0583	·1189	·2120
	10	·0000	·0000	·0000	·0000	·0003	·0017	·0060	·0175	·0426	·0898
	11	·0000	·0000	·0000	·0000	·0000	·0002	·0011	·0039	·0114	·0287
	12	·0000	·0000	·0000	·0000	·0000	·0000	·0001	·0006	·0022	·0065
	13	·0000	·0000	·0000	·0000	·0000	·0000	·0000	·0001	·0003	·0009
	14	·0000	·0000	·0000	·0000	·0000	·0000	·0000	·0000	·0000	·0001

TABLE D3 (*continued*)

n	s	·05	·10	·15	·20	·25	·30	·35	·40	·45	·50
15	1	·5367	·7941	·9126	·9648	·9866	·9953	·9984	·9995	·9999	1·0000
	2	·1710	·4510	·6814	·8329	·9198	·9647	·9858	·9948	·9983	·9995
	3	·0362	·1841	·3958	·6020	·7639	·8732	·9383	·9729	·9893	·9963
	4	·0055	·0556	·1773	·3518	·5387	·7031	·8273	·9095	·9576	·9824
	5	·0006	·0127	·0617	·1642	·3135	·4845	·6481	·7827	·8796	·9408
	6	·0001	·0022	·0168	·0611	·1484	·2784	·4357	·5968	·7392	·8491
	7	·0000	·0003	·0036	·0181	·0566	·1311	·2452	·3902	·5478	·6964
	8	·0000	·0000	·0006	·0042	·0173	·0500	·1132	·2131	·3465	·5000
	9	·0000	·0000	·0001	·0008	·0042	·0152	·0422	·0950	·1818	·3036
	10	·0000	·0000	·0000	·0001	·0008	·0037	·0124	·0338	·0769	·1509
	11	·0000	·0000	·0000	·0000	·0001	·0007	·0028	·0093	·0255	·0592
	12	·0000	·0000	·0000	·0000	·0000	·0001	·0005	·0019	·0063	·0176
	13	·0000	·0000	·0000	·0000	·0000	·0000	·0001	·0003	·0011	·0037
	14	·0000	·0000	·0000	·0000	·0000	·0000	·0000	·0000	·0001	·0005
	15	·0000	·0000	·0000	·0000	·0000	·0000	·0000	·0000	·0000	·0000
16	1	·5599	·8147	·9257	·9719	·9900	·9967	·9990	·9997	·9999	1·0000
	2	·1892	·4853	·7161	·8593	·9365	·9739	·9902	·9967	·9990	·9997
	3	·0429	·2108	·4386	·6482	·8029	·9006	·9549	·9817	·9934	·9979
	4	·0070	·0684	·2101	·4019	·5950	·7541	·8661	·9349	·9719	·9894
	5	·0009	·0170	·0791	·2018	·3698	·5501	·7108	·8334	·9147	·9616
	6	·0001	·0033	·0235	·0817	·1897	·3402	·5100	·6712	·8024	·8949
	7	·0000	·0005	·0056	·0267	·0796	·1753	·3119	·4728	·6340	·7228
	8	·0000	·0001	·0011	·0070	·0271	·0744	·1594	·2839	·4371	·5982
	9	·0000	·0000	·0002	·0015	·0075	·0257	·0671	·1423	·2559	·4018
	10	·0000	·0000	·0000	·0002	·0016	·0071	·0229	·0583	·1241	·2272
	11	·0000	·0000	·0000	·0000	·0003	·0016	·0062	·0191	·0486	·1051
	12	·0000	·0000	·0000	·0000	·0000	·0003	·0013	·0049	·0149	·0384
	13	·0000	·0000	·0000	·0000	·0000	·0000	·0002	·0009	·0035	·0106
	14	·0000	·0000	·0000	·0000	·0000	·0000	·0000	·0001	·0006	·0021
	15	·0000	·0000	·0000	·0000	·0000	·0000	·0000	·0000	·0001	·0003
	16	·0000	·0000	·0000	·0000	·0000	·0000	·0000	·0000	·0000	·0000
17	1	·5189	·8332	·9369	·9775	·9925	·9977	·9993	·9998	1·0000	1·0000
	2	·2078	·5182	·7475	·8818	·9499	·9807	·9933	·9979	·9994	·9999
	3	·0503	·2382	·4802	·6904	·8363	·9226	·9673	·9877	·9959	·9988
	4	·0088	·0826	·2444	·4511	·6470	·7981	·8972	·9536	·9816	·9936
	5	·0012	·0221	·0987	·2418	·4261	·6113	·7652	·8740	·9404	·9755
	6	·0001	·0047	·0319	·1057	·2347	·4032	·5803	·7361	·8529	·9283
	7	·0000	·0008	·0083	·0377	·1071	·2248	·3812	·5522	·7098	·8338
	8	·0000	·0001	·0017	·0109	·0402	·1046	·2128	·3595	·5257	·6855
	9	·0000	·0000	·0003	·0026	·0124	·0403	·0994	·1989	·3374	·5000
	10	·0000	·0000	·0000	·0005	·0031	·0127	·0383	·0919	·1834	·3145

TABLE D3 (*continued*)

n	s	·05	·10	·15	·20	·25	·30	·35	·40	·45	·50
	11	·0000	·0000	·0000	·0001	·0006	·0032	·0120	·0348	·0826	·1662
	12	·0000	·0000	·0000	·0000	·0001	·0007	·0030	·0106	·0301	·0717
	13	·0000	·0000	·0000	·0000	·0000	·0001	·0006	·0025	·0086	·0245
	14	·0000	·0000	·0000	·0000	·0000	·0000	·0000	·0005	·0019	·0064
	15	·0000	·0000	·0000	·0000	·0000	·0000	·0000	·0001	·0003	·0012
	16	·0000	·0000	·0000	·0000	·0000	·0000	·0000	·0000	·0000	·0001
	17	·0000	·0000	·0000	·0000	·0000	·0000	·0000	·0000	·0000	·0000
18	1	·6028	·8499	·9464	·9820	·9944	·9984	·9996	·9999	1·0000	1·0000
	2	·2265	·5497	·7759	·9009	·9605	·9858	·9954	·9987	·9997	·9999
	3	·0581	·2662	·5203	·7287	·8647	·9400	·9764	·9918	·9975	·9993
	4	·0109	·0982	·2798	·4990	·6943	·8354	·9217	·9672	·9880	·9962
	5	·0015	·0282	·1206	·2836	·4813	·6673	·8114	·9058	·9589	·9846
	6	·0002	·0064	·0419	·1329	·2825	·4656	·6450	·7912	·8923	·9519
	7	·0000	·0012	·0118	·0513	·1390	·2783	·4509	·6257	·7742	·8811
	8	·0000	·0002	·0027	·0163	·0569	·1407	·2717	·4366	·6085	·7597
	9	·0000	·0000	·0005	·0043	·0193	·0596	·1391	·2632	·4222	·5927
	10	·0000	·0000	·0001	·0009	·0054	·0210	·0597	·1347	·2527	·4073
	11	·0000	·0000	·0000	·0002	·0012	·0061	·0212	·0576	·1280	·2403
	12	·0000	·0000	·0000	·0000	·0002	·0014	·0062	·0203	·0537	·1189
	13	·0000	·0000	·0000	·0000	·0000	·0003	·0014	·0058	·0183	·0481
	14	·0000	·0000	·0000	·0000	·0000	·0000	·0003	·0013	·0049	·0154
	15	·0000	·0000	·0000	·0000	·0000	·0000	·0000	·0002	·0010	·0038
	16	·0000	·0000	·0000	·0000	·0000	·0000	·0000	·0000	·0001	·0007
	17	·0000	·0000	·0000	·0000	·0000	·0000	·0000	·0000	·0000	·0001
	18	·0000	·0000	·0000	·0000	·0000	·0000	·0000	·0000	·0000	·0000
19	1	·6226	·8649	·9544	·9856	·9958	·9989	·9997	·9999	1·0000	1·0000
	2	·2453	·5797	·8015	·9171	·9690	·9896	·9969	·9992	·9998	1·0000
	3	·0665	·2946	·5587	·7631	·8887	·9538	·9830	·9945	·9985	·9996
	4	·0132	·1150	·3159	·5449	·7639	·8668	·9409	·9770	·9923	·9978
	5	·0020	·0352	·1444	·3267	·5346	·7178	·8500	·9304	·9720	·9904
	6	·0002	·0086	·0537	·1631	·3322	·5261	·7032	·8371	·9223	·9682
	7	·0000	·0017	·0163	·0676	·1749	·3345	·5188	·6919	·8273	·9165
	8	·0000	·0003	·0041	·0233	·0775	·1820	·3344	·5122	·6831	·8204
	9	·0000	·0000	·0008	·0067	·0287	·0839	·1855	·3325	·5060	·6762
	10	·0000	·0000	·0001	·0016	·0089	·0326	·0875	·1861	·3290	·5000
	11	·0000	·0000	·0000	·0003	·0023	·0105	·0347	·0885	·1841	·3238
	12	·0000	·0000	·0000	·0000	·0005	·0028	·0114	·0352	·0871	·1796
	13	·0000	·0000	·0000	·0000	·0001	·0006	·0031	·0116	·0342	·0835
	14	·0000	·0000	·0000	·0000	·0000	·0001	·0007	·0031	·0109	·0318
	15	·0000	·0000	·0000	·0000	·0000	·0000	·0001	·0006	·0028	·0096
	16	·0000	·0000	·0000	·0000	·0000	·0000	·0000	·0001	·0005	·0022
	17	·0000	·0000	·0000	·0000	·0000	·0000	·0000	·0000	·0001	·0004

TABLE D3 (*continued*)

n	s	·05	·10	·15	·20	·25	·30	·35	·40	·45	·50
	18	·0000	·0000	·0000	·0000	·0000	·0000	·0000	·0000	·0000	·0000
	19	·0000	·0000	·0000	·0000	·0000	·0000	·0000	·0000	·0000	·0000
20	1	·6415	·8784	·9612	·9885	·9968	·9992	·9998	1·0000	1·0000	1·0000
	2	·2642	·6083	·8244	·9308	·9757	·9924	·9979	·9995	·9999	1·0000
	3	·0755	·3231	·5951	·7939	·9087	·9645	·9879	·9964	·9991	·9998
	4	·0159	·1330	·3523	·5886	·7748	·8929	·9556	·9840	·9951	·9987
	5	·0026	·0432	·1702	·3704	·5852	·7625	·8818	·9490	·9811	·9941
	6	·0003	·0113	·0673	·1958	·3828	·5836	·7546	·8744	·9447	·9793
	7	·0000	·0024	·0219	·0867	·2142	·3920	·5834	·7500	·8701	·9423
	8	·0000	·0004	·0059	·0321	·1018	·2277	·3990	·5841	·7480	·8684
	9	·0000	·0001	·0013	·0100	·0409	·1133	·2376	·4044	·5857	·7483
	10	·0000	·0000	·0002	·0026	·0139	·0480	·1218	·2447	·4086	·5881
	11	·0000	·0000	·0000	·0006	·0039	·0171	·0532	·1275	·2493	·4119
	12	·0000	·0000	·0000	·0001	·0009	·0051	·0196	·0565	·1308	·2517
	13	·0000	·0000	·0000	·0000	·0002	·0013	·0060	·0210	·0580	·1316
	14	·0000	·0000	·0000	·0000	·0000	·0008	·0015	·0065	·0214	·0577
	15	·0000	·0000	·0000	·0000	·0000	·0000	·0003	·0016	·0064	·0207
	16	·0000	·0000	·0000	·0000	·0000	·0000	·0000	·0003	·0015	·0059
	17	·0000	·0000	·0000	·0000	·0000	·0000	·0000	·0000	·0003	·0013
	18	·0000	·0000	·0000	·0000	·0000	·0000	·0000	·0000	·0000	·0002
	19	·0000	·0000	·0000	·0000	·0000	·0000	·0000	·0000	·0000	·0000
	20	·0000	·0000	·0000	·0000	·0000	·0000	·0000	·0000	·0000	·0000

Table D4

The Poisson density function is given by

$$f(x) = \frac{e^{-m}\, m^x}{x!}, \qquad m > 0, \quad x = 0, 1, 2, \ldots,$$

This table gives values of

$$F = \sum_{x=x'}^{\infty} \frac{e^{-m}\, m^x}{x!},$$

for specified values of x' and m.

x'	m 0·1	0·2	0·3	0·4	0·5	0·6	0·7	0·8	0·9	1·0
0	1·0000	1·0000	1·0000	1·0000	1·0000	1·0000	1·0000	1·0000	1·0000	1·0000
1	·0952	·1813	·2592	·3297	·3935	·4512	·5034	·5507	·5934	·6321
2	·0047	·0175	·0369	·0616	·0902	·1219	·1558	·1912	·2275	·2642
3	·0002	·0011	·0036	·0079	·0144	·0231	·0341	·0474	·0629	·0803
4	·0000	·0001	·0003	·0008	·0018	·0034	·0058	·0091	·0135	·0190
5	·0000	·0000	·0000	·0001	·0002	·0004	·0008	·0014	·0023	·0037
6	·0000	·0000	·0000	·0000	·0000	·0000	·0001	·0002	·0003	·0006
7	·0000	·0000	·0000	·0000	·0000	·0000	·0000	·0000	·0000	·0001

x'	1·1	1·2	1·3	1·4	1·5	1·6	1·7	1·8	1·9	2·0
0	1·0000	1·0000	1·0000	1·0000	1·0000	1·0000	1·0000	1·0000	1·0000	1·0000
1	·6671	·6988	·7275	·7534	·7769	·7981	·8173	·8347	·8504	·8647
2	·3010	·3374	·3732	·4082	·4422	·4751	·5068	·5372	·5663	·5940
3	·0996	·1205	·1429	·1665	·1912	·2166	·2428	·2694	·2963	·3233
4	·0257	·0338	·0431	·0537	·0656	·0788	·0932	·1087	·1253	·1429
5	·0054	·0077	·0107	·0143	·0186	·0237	·0296	·0364	·0441	·0527
6	·0010	·0015	·0022	·0032	·0045	·0060	·0080	·0104	·0132	·0166
7	·0001	·0003	·0004	·0006	·0009	·0013	·0019	·0026	·0034	·0045
8	·0000	·0000	·0001	·0001	·0002	·0003	·0004	·0006	·0008	·0011
9	·0000	·0000	·0000	·0000	·0000	·0000	·0001	·0001	·0002	·0002

x'	2·1	2·2	2·3	2·4	2·5	2·6	2·7	2·8	2·9	3·0
0	1·0000	1·0000	1·0000	1·0000	1·0000	1·0000	1·0000	1·0000	1·0000	1·0000
1	·8775	·8892	·8997	·9093	·9179	·9257	·9328	·9392	·9450	·9502
2	·6204	·6454	·6691	·6916	·7127	·7326	·7513	·7689	·7854	·8009
3	·3504	·3773	·4040	·4303	·4562	·4816	·5064	·5305	·5540	·5768
4	·1614	·1806	·2007	·2213	·2424	·2640	·2859	·3081	·3304	·3528
5	·0621	·0725	·0838	·0959	·1088	·1226	·1371	·1523	·1682	·1847
6	·0204	·0249	·0300	·0357	·0420	·0490	·0567	·0651	·0742	·0839
7	·0059	·0075	·0094	·0116	·0142	·0172	·0206	·0244	·0287	·0335
8	·0015	·0020	·0026	·0033	·0042	·0053	·0066	·0081	·0099	·0119
9	·0003	·0005	·0006	·0009	·0011	·0015	·0019	·0024	·0031	·0038
10	·0001	·0001	·0001	·0002	·0003	·0004	·0005	·0007	·0009	·0011
11	·0000	·0000	·0000	·0000	·0001	·0001	·0001	·0002	·0002	·0003
12	·0000	·0000	·0000	·0000	·0000	·0000	·0000	·0000	·0001	·0001

TABLE D4 (*continued*)

x'	m 3·1	3·2	3·3	3·4	3·5	3·6	3·7	3·8	3·9	4·0
0	1·0000	1·0000	1·0000	1·0000	1·0000	1·0000	1·0000	1·0000	1·0000	1·0000
1	·9550	·9592	·9631	·9666	·9698	·9727	·9753	·9776	·9798	·9817
2	·8153	·8288	·8414	·8532	·8641	·8743	·8838	·8926	·9008	·9084
3	·5988	·6201	·6406	·6603	·6792	·6973	·7146	·7311	·7469	·7619
4	·3752	·3975	·4197	·4416	·4634	·4848	·5058	·5265	·5468	·5665
5	·2018	·2194	·2374	·2558	·2746	·2936	·3128	·3322	·3516	·3712
6	·0943	·1054	·1171	·1295	·1424	·1559	·1699	·1844	·1994	·2149
7	·0388	·0446	·0510	·0579	·0653	·0732	·0818	·0919	·1005	·1107
8	·0142	·0168	·0198	·0231	·0267	·0308	·0352	·0401	·0454	·0511
9	·0047	·0057	·0069	·0083	·0099	·0117	·0137	·0160	·0185	·0214
10	·0014	·0018	·0022	·0027	·0033	·0040	·0048	·0058	·0069	·0081
11	·0004	·0005	·0006	·0008	·0010	·0013	·0016	·0019	·0023	·0028
12	·0001	·0001	·0002	·0002	·0003	·0004	·0005	·0006	·0007	·0009
13	·0000	·0000	·0000	·0001	·0001	·0001	·0001	·0002	·0002	·0003
14	·0000	·0000	·0000	·0000	·0000	·0000	·0000	·0000	·0001	·0001

x'	4·1	4·2	4·3	4·4	4·5	4·6	4·7	4·8	4·9	5·0
0	1·0000	1·0000	1·0000	1·0000	1·0000	1·0000	1·0000	1·0000	1·0000	1·0000
1	·9834	·9850	·9864	·9877	·9889	·9899	·9909	·9918	·9926	·9933
2	·9155	·9220	·9281	·9337	·9389	·9437	·9482	·9523	·9561	·9596
3	·7762	·7898	·8026	·8149	·8264	·8374	·8477	·8575	·8667	·8753
4	·5858	·6046	·6228	·6406	·6577	·6743	·6903	·7058	·7207	·7350
5	·3907	·4102	·4296	·4488	·4679	·4868	·5054	·5237	·5418	·5595
6	·2307	·2469	·2633	·2801	·2971	·3412	·3316	·3490	·3665	·3840
7	·1214	·1325	·1442	·1564	·1689	·1820	·1954	·2092	·2233	·2378
8	·0573	·0639	·0710	·0786	·0866	·0951	·1040	·1133	·1231	·1334
9	·0245	·0279	·0317	·0358	·0403	·0451	·0503	·0558	·0618	·0681
10	·0095	·0111	·0129	·0149	·0171	·0195	·0222	·0251	·0283	·0318
11	·0034	·0041	·0048	·0057	·0067	·0078	·0090	·0104	·0120	·0137
12	·0011	·0014	·0017	·0020	·0024	·0029	·0034	·0040	·0047	·0055
13	·0003	·0004	·0005	·0007	·0008	·0010	·0012	·0014	·0017	·0020
14	·0001	·0001	·0002	·0002	·0003	·0003	·0004	·0005	·0006	·0007
15	·0000	·0000	·0000	·0001	·0001	·0001	·0001	·0001	·0002	·0002
16	·0000	·0000	·0000	·0000	·0000	·0000	·0000	·0000	·0001	·0001

TABLE D4 (*continued*)

	m									
x′	5·1	5·2	5·3	5·4	5·5	5·6	5·7	5·8	5·9	6·0
0	1·0000	1·0000	1·0000	1·0000	1·0000	1·0000	1·0000	1·0000	1·0000	1·0000
1	·9939	·9945	·9950	·9955	·9959	·9963	·9967	·9970	·9973	·9975
2	·9628	·9658	·9686	·9711	·9734	·9756	·9776	·9794	·9811	·9826
3	·8835	·8912	·8984	·9052	·9116	·9176	·9232	·9285	·9334	·9380
4	·7487	·7619	·7746	·7867	·7983	·8094	·8200	·8300	·8396	·8488
5	·5769	·5939	·6105	·6267	·6425	·6579	·6728	·6873	·7013	·7149
6	·4016	·4191	·4365	·4539	·4711	·4881	·5050	·5217	·5381	·5543
7	·2526	·2676	·2829	·2983	·3140	·3297	·3456	·3616	·3776	·3937
8	·1440	·1551	·1665	·1783	·1905	·2030	·2159	·2290	·2424	·2560
9	·0748	·0819	·0894	·0974	·1056	·1143	·1234	·1328	·1426	·1528
10	·0356	·0397	·0441	·0488	·0538	·0591	·0648	·0708	·0722	·0839
11	·0156	·0177	·0200	·0225	·0253	·0282	·0314	·0349	·0386	·0426
12	·0063	·0073	·0084	·0096	·0110	·0125	·0141	·0160	·0179	·0201
13	·0024	·0028	·0033	·0038	·0045	·0051	·0059	·0068	·0078	·0088
14	·0008	·0010	·0012	·0014	·0017	·0020	·0023	·0027	·0031	·0036
15	·0003	·0003	·0004	·0005	·0006	·0007	·0009	·0010	·0012	·0014
16	·0001	·0001	·0001	·0002	·0002	·0002	·0003	·0004	·0004	·0005
17	·0000	·0000	·0000	·0001	·0001	·0001	·0001	·0001	·0001	·0002
18	·0000	·0000	·0000	·0000	·0000	·0000	·0000	·0000	·0000	·0001

x′	6·1	6·2	6·3	6·4	6·5	6·6	6·7	6·8	6·9	7·0
0	1·0000	1·0000	1·0000	1·0000	1·0000	1·0000	1·0000	1·0000	1·0000	1·0000
1	·9978	·9980	·9982	·9983	·9985	·9986	·9988	·9989	·9990	·9991
2	·9841	·9854	·9866	·9877	·9887	·9897	·9905	·9913	·9920	·9927
3	·9423	·9464	·9502	·9537	·9570	·9600	·9629	·9656	·9680	·9704
4	·8575	·8658	·8736	·8811	·8882	·8948	·9012	·9072	·9129	·9182
5	·7281	·7408	·7531	·7649	·7763	·7873	·7978	·8080	·8177	·8270
6	·5702	·5859	·6012	·6163	·6310	·6453	·6594	·6730	·6863	·6993
7	·4098	·4258	·4418	·4577	·4735	·4892	·5047	·5201	·5353	·5503
8	·2699	·2840	·2983	·3127	·3272	·3419	·3567	·3715	·3864	·4013
9	·1633	·1741	·1852	·1967	·2084	·2204	·2327	·2452	·2580	·2709
10	·0910	·0984	·1061	·1142	·1226	·1314	·1404	·1498	·1505	·1695
11	·0469	·0514	·0563	·0614	·0668	·0726	·0786	·0849	·0916	·0985
12	·0224	·0250	·0277	·0307	·0339	·0373	·0409	·0448	·0495	·0534
13	·0100	·0113	·0127	·0143	·0160	·0179	·0199	·0221	·0245	·0270
14	·0042	·0048	·0055	·0063	·0071	·0080	·0091	·0102	·0115	·0128
15	·0016	·0019	·0022	·0026	·0030	·0034	·0039	·0044	·0050	·0057
16	·0006	·0007	·0008	·0010	·0012	·0014	·0016	·0018	·0021	·0024
17	·0002	·0003	·0003	·0004	·0004	·0005	·0006	·0007	·0008	·0010
18	·0001	·0001	·0001	·0001	·0002	·0002	·0002	·0003	·0003	·0004
19	·0000	·0000	·0000	·0000	·0001	·0001	·0001	·0001	·0001	·0001

TABLE D4 (*continued*)

x'	7·1	7·2	7·3	7·4	7·5	7·6	7·7	7·8	7·9	8·0
0	1·0000	1·0000	1·0000	1·0000	1·0000	1·0000	1·0000	1·0000	1·0000	1·0000
1	·9992	·9993	·9993	·9994	·9994	·9995	·9995	·9996	·9996	·9997
2	·9933	·9939	·9944	·9949	·9953	·9957	·9961	·9964	·9967	·9970
3	·9725	·9745	·9764	·9781	·9797	·9812	·9826	·9839	·9851	·9862
4	·9233	·9281	·9326	·9368	·9409	·9446	·9482	·9515	·9547	·9576
5	·8359	·8445	·8527	·8605	·8679	·8751	·8819	·8883	·8945	·9004
6	·7119	·7241	·7360	·7474	·7586	·7693	·7797	·7897	·7994	·8088
7	·5651	·5796	·5940	·6080	·6218	·6354	·6486	·6616	·6743	·6866
8	·4162	·4311	·4459	·4607	·4754	·4900	·5044	·5188	·5330	·5470
9	·2840	·2973	·3108	·3243	·3380	·3518	·3657	·3796	·3935	·4075
10	·1798	·1904	·2012	·2123	·2236	·2351	·2469	·2589	·2710	·2834
11	·1058	·1133	·1212	·1293	·1378	·1465	·1555	·1648	·1743	·1841
12	·0580	·0629	·0681	·0735	·0792	·0852	·0915	·0980	·1048	·1119
13	·0297	·0327	·0358	·0391	·0427	·0464	·0504	·0546	·0591	·0638
14	·0143	·0159	·0176	·0195	·0216	·0238	·0261	·0286	·0313	·0342
15	·0065	·0073	·0082	·0092	·0103	·0114	·0127	·0141	·0156	·0173
16	·0028	·0031	·0036	·0041	·0046	·0052	·0059	·0066	·0074	·0082
17	·0011	·0013	·0015	·0017	·0020	·0022	·0026	·0029	·0033	·0037
18	·0004	·0005	·0006	·0007	·0008	·0009	·0011	·0012	·0014	·0016
19	·0002	·0002	·0002	·0003	·0003	·0004	·0004	·0005	·0006	·0005
20	·0001	·0001	·0001	·0001	·0001	·0001	·0002	·0002	·0002	·0003
21	·0000	·0000	·0000	·0000	·0000	·0000	·0001	·0001	·0001	·0001

x'	8·1	8·2	8·3	8·4	8·5	8·6	8·7	8·8	8·9	9·0
0	1·0000	1·0000	1·0000	1·0000	1·0000	1·0000	1·0000	1·0000	1·0000	1·0000
1	·9997	·9997	·9998	·9998	·9998	·9998	·9998	·9998	·9999	·9999
2	·9972	·9975	·9977	·9979	·9981	·9982	·9984	·9985	·9987	·9988
3	·9873	·9882	·9891	·9900	·9907	·9914	·9921	·9927	·9932	·9938
4	·9604	·9630	·9654	·9677	·9699	·9719	·9738	·9756	·9772	·9788
5	·9060	·9113	·9163	·9211	·9256	·9299	·9340	·9379	·9416	·9450
6	·8178	·8264	·8347	·8427	·8504	·8578	·8648	·8716	·8781	·8843
7	·6987	·7104	·7219	·7330	·7438	·7543	·7645	·7744	·7840	·7932
8	·5609	·5746	·5881	·6013	·6144	·6272	·6398	·6522	·6643	·6761
9	·4214	·4353	·4493	·4631	·4769	·4906	·5042	·5177	·5311	·5443
10	·2959	·3085	·3212	·3341	·3470	·3600	·3731	·3863	·3994	·4126
11	·1942	·2045	·2150	·2257	·2366	·2478	·2591	·2706	·2822	·2940
12	·1193	·1269	·1348	·1429	·1513	·1600	·1689	·1780	·1874	·1970
13	·0687	·0739	·0793	·0850	·0909	·0971	·1035	·1102	·1171	·1242
14	·0372	·0405	·0439	·0476	·0514	·0555	·0597	·0642	·0689	·0739
15	·0190	·0209	·0229	·0251	·0274	·0299	·0325	·0353	·0383	·0415
16	·0092	·0102	·0113	·0125	·0138	·0152	·0168	·0184	·0202	·0220
17	·0042	·0047	·0053	·0059	·0066	·0074	·0082	·0091	·0101	·0111
18	·0018	·0021	·0023	·0027	·0030	·0034	·0038	·0043	·0048	·0053
19	·0008	·0009	·0010	·0011	·0013	·0015	·0017	·0019	·0022	·0024
20	·0003	·0003	·0004	·0005	·0005	·0006	·0007	·0008	·0009	·0011
21	·0001	·0001	·0002	·0002	·0002	·0002	·0003	·0003	·0004	·0004
22	·0000	·0000	·0001	·0001	·0001	·0001	·0001	·0001	·0002	·0002
23	·0000	·0000	·0000	·0000	·0000	·0000	·0000	·0000	·0001	·0001

TABLE D4 (*continued*)

x'	m 9·1	9·2	9·3	9·4	9·5	9·6	9·7	9·8	9·9	10
0	1·0000	1·0000	1·0000	1·0000	1·0000	1·0000	1·0000	1·0000	1·0000	1·0000
1	·9999	·9999	·9999	·9999	·9999	·9999	·9999	·9999	1·0000	1·0000
2	·9989	·9990	·9991	·9991	·9992	·9993	·9993	·9994	·9995	·9995
3	·9942	·9947	·9951	·9955	·9958	·9962	·9965	·9967	·9970	·9972
4	·9802	·9816	·9828	·9840	·9851	·9862	·9871	·9880	·9889	·9897
5	·9483	·9514	·9544	·9571	·9597	·9622	·9645	·9667	·9688	·9707
6	·8902	·8959	·9014	·9065	·9115	·9162	·9207	·9250	·9290	·9329
7	·8022	·8108	·8192	·8273	·8351	·8426	·8498	·8567	·8634	·8699
8	·6877	·6990	·7101	·7208	·7313	·7416	·7515	·7612	·7706	·7798
9	·5574	·5704	·5832	·5958	·6082	·6204	·6324	·6442	·6558	·6672
10	·4258	·4389	·4521	·4651	·4782	·4911	·5040	·5168	·5295	·5421
11	·3059	·3180	·3301	·3424	·3547	·3671	·3795	·3920	·4045	·4170
12	·2068	·2168	·2270	·2374	·2480	·2588	·2697	·2807	·2919	·3032
13	·1316	·1393	·1471	·1552	·1636	·1721	·1809	·1899	·1991	·2084
14	·0790	·0844	·0900	·0958	·1019	·1081	·1147	·1214	·1284	·1355
15	·0448	·0483	·0520	·0559	·0600	·0643	·0688	·0735	·0784	·0835
16	·0240	·0262	·0285	·0309	·0335	·0362	·0391	·0421	·0454	·0487
17	·0122	·0135	·0148	·0162	·0177	·0194	·0211	·0230	·0249	·0270
18	·0059	·0066	·0073	·0081	·0089	·0098	·0108	·0119	·0130	·0143
19	·0027	·0031	·0034	·0038	·0043	·0048	·0053	·0059	·0065	·0072
20	·0012	·0014	·0015	·0017	·0020	·0022	·0025	·0028	·0031	·0035
21	·0005	·0006	·0007	·0008	·0009	·0010	·0011	·0013	·0014	·0016
22	·0002	·0002	·0003	·0003	·0004	·0004	·0005	·0005	·0006	·0007
23	·0001	·0001	·0001	·0001	·0001	·0002	·0002	·0002	·0003	·0003
24	·0000	·0000	·0000	·0000	·0001	·0001	·0001	·0001	·0001	·0001

TABLE D5 This table gives values of χ^2 for specified values of ν and F where $F(\chi^2, \nu) = \frac{1}{2^{\nu/2}\,\Gamma(\nu/2)} \int_0^{\chi^2} dt\, t^{\nu/2-1}\, e^{-t/2}$.

ν	F ·005	·010	·025	·050	·100	·250	·500	·750	·900	·950	·975	·990	·995
1	·0000393	·000157	·000982	·00393	·0158	·102	·455	1·32	2·71	3·84	5·02	6·63	7·88
2	·0100	·0201	·0506	·103	·211	·575	1·39	2·77	4·61	5·99	7·38	9·21	10·6
3	·0717	·115	·216	·352	·584	1·21	2·37	4·11	6·25	7·81	9·35	11·3	12·8
4	·207	·297	·484	·711	1·06	1·92	3·36	5·39	7·78	9·49	11·1	13·3	14·9
5	·412	·554	·831	1·15	1·61	2·67	4·35	6·63	9·24	11·1	12·8	15·1	16·7
6	·676	·872	1·24	1·64	2·20	3·45	5·35	7·84	10·6	12·6	14·4	16·8	18·5
7	·989	1·24	1·69	2·17	2·83	4·25	6·35	9·04	12·0	14·1	16·0	18·5	20·3
8	1·34	1·65	2·18	2·73	3·49	5·07	7·34	10·2	13·4	15·5	17·5	20·1	22·0
9	1·73	2·09	2·70	3·33	4·17	5·90	8·34	11·4	14·7	16·9	19·0	21·7	23·6
10	2·16	2·56	3·25	3·94	4·87	6·74	9·34	12·5	16·0	18·3	20·5	23·2	25·2
11	2·60	3·05	3·82	4·57	5·58	7·58	10·3	13·7	17·3	19·7	21·9	24·7	26·8
12	3·07	3·57	4·40	5·23	6·30	8·44	11·3	14·8	18·5	21·0	23·3	26·2	28·3
13	3·57	4·11	5·01	5·89	7·04	9·30	12·3	16·0	19·8	22·4	24·7	27·7	29·8
14	4·07	4·66	5·63	6·57	7·79	10·2	13·3	17·1	21·1	23·7	26·1	29·1	31·3
15	4·60	5·23	6·26	7·26	8·55	11·0	14·3	18·2	22·3	25·0	27·5	30·6	32·8
16	5·14	5·81	6·91	7·96	9·31	11·9	15·3	19·4	23·5	26·3	28·8	32·0	34·3
17	5·70	6·41	7·56	8·67	10·1	12·8	16·3	20·5	24·8	27·6	30·2	33·4	35·7
18	6·26	7·01	8·23	9·39	10·9	13·7	17·3	21·6	26·0	28·9	31·5	34·8	37·2
19	6·84	7·63	8·91	10·1	11·7	14·6	18·3	22·7	27·2	30·1	32·9	36·2	38·6
20	7·43	8·26	9·59	10·9	12·4	15·5	19·3	23·8	28·4	31·4	34·2	37·6	40·0
21	8·03	8·90	10·3	11·3	13·2	16·3	20·3	24·9	29·6	32·7	35·5	38·9	41·4
22	8·64	9·54	11·0	12·3	14·0	17·2	21·3	26·0	30·8	33·9	36·8	40·3	42·8
23	9·26	10·2	11·7	13·1	14·8	18·1	22·3	27·1	32·0	35·2	38·1	41·6	44·2
24	9·89	10·9	12·4	13·8	15·7	19·0	23·3	28·2	33·2	36·4	39·4	43·0	45·6
25	10·5	11·5	13·1	14·6	16·5	19·9	24·3	29·3	34·4	37·7	40·6	44·3	46·9
26	11·2	12·2	13·8	15·4	17·3	20·8	25·3	30·4	35·6	38·9	41·9	45·6	48·3
27	11·8	12·9	14·6	16·2	18·1	21·7	26·3	31·5	36·7	40·1	43·2	47·0	49·6
28	12·5	13·6	15·3	16·9	18·9	22·7	27·3	32·6	37·9	41·3	44·5	48·3	51·0
29	13·1	14·3	16·0	17·7	19·8	23·6	28·3	33·7	39·1	42·6	45·7	49·6	52·3
30	13·8	15·0	16·8	18·5	20·6	24·5	29·3	34·8	40·3	43·8	47·0	50·9	53·7

TABLE D6

This table gives values of t for specified values of ν and F where

$$F(t;\nu) = \frac{1}{(\pi\nu)^{\frac{1}{2}}} \frac{\Gamma[(\nu+1)/2]}{\Gamma(\nu/2)} \int_{-\infty}^{t} dx \left(1 + \frac{x^2}{\nu}\right)^{-(\nu+1)/2}.$$

Note that

$$F(-t) = 1 - F(t).$$

	F							
ν	·60	·75	·90	·95	·975	·99	·995	·9995
1	·325	1·000	3·078	6·314	12·706	31·821	63·657	636·619
2	·289	·816	1·886	2·920	4·303	6·695	9·925	31·598
3	·277	·765	1·638	2·353	3·182	4·541	5·841	12·941
4	·271	·741	1·533	2·132	2·776	3·747	4·604	8·610
5	·267	·727	1·476	2·015	2·571	3·365	4·032	6·859
6	·265	·718	1·440	1·943	2·447	3·143	3·707	5·959
7	·263	·711	1·415	1·895	2·365	2·998	3·499	5·405
8	·262	·706	1·397	1·860	2·306	2·896	3·355	5·041
9	·261	·703	1·383	1·833	2·262	2·821	3·250	4·781
10	·260	·700	1·372	1·812	2·228	2·764	3·169	4·587
11	·260	·697	1·363	1·796	2·201	2·718	3·106	4·437
12	·259	·695	1·356	1·782	2·179	2·681	3·055	4·318
13	·259	·694	1·350	1·771	2·160	2·650	3·012	4·221
14	·258	·692	1·345	1·761	2·145	2·624	2·977	4·140
15	·258	·691	1·341	1·753	2·131	2·602	2·947	4·073
16	·258	·690	1·337	1·746	2·120	2·583	2·921	4·015
17	·257	·689	1·333	1·740	2·110	2·567	2·898	3·965
18	·257	·688	1·330	1·734	2·101	2·552	2·878	3·922
19	·257	·688	1·328	1·729	2·093	2·539	2·861	3·883
20	·257	·687	1·325	1·725	2·086	2·528	2·845	3·850
21	·257	·686	1·323	1·721	2·080	2·518	2·831	3·819
22	·256	·686	1·321	1·717	2·074	2·508	2·819	3·792
23	·256	·685	1·319	1·714	2·069	2·500	2·807	3·767
24	·256	·685	1·318	1·711	2·064	2·492	2·797	3·745
25	·256	·684	1·316	1·708	2·060	2·485	2·787	3·725
26	·256	·684	1·315	1·706	2·056	2·479	2·779	3·707
27	·256	·684	1·314	1·703	2·052	2·473	2·771	3·690
28	·256	·683	1·313	1·701	2·048	2·467	2·763	3·674
29	·256	·683	1·311	1·699	2·045	2·462	2·756	3·659
30	·256	·683	1·310	1·697	2·042	2·457	2·750	3·646
40	·255	·681	1·303	1·684	2·021	2·423	2·704	3·551
60	·254	·679	1·296	1·671	2·000	2·390	2·660	3·460
120	·254	·677	1·289	1·658	1·980	2·358	2·617	3·373
∞	·253	·674	1·282	1·645	1·960	2·326	2·576	3·291

TABLE D7

This table gives values of F such that

$$F(F; m, n) = \frac{\Gamma[(m+n)/2]}{\Gamma(m/2)\,\Gamma(n/2)} \left(\frac{m}{n}\right)^{m/2} \int_0^F dx \frac{x^{(m-2)/2}}{[1+(m/n)\,x]^{(m+n)/2}},$$

for specified values of m and n. It can also be used for values corresponding to $F(F) = 0{\cdot}10$, $0{\cdot}05$, $0{\cdot}025$, $0{\cdot}01$, $0{\cdot}005$ and $0{\cdot}001$ by use of the relation

$$F_{1-\alpha}(n, m) = [F_\alpha(m, n)]^{-1}.$$

$F(F; m, n) = 0{\cdot}90$

	m								
n	1	2	3	4	5	6	7	8	9
1	39·86	49·50	53·59	55·83	57·24	58·20	58·81	59·44	59·86
2	8·53	9·00	9·16	9·24	9·29	9·33	9·35	9·37	9·38
3	5·54	5·46	5·39	5·34	5·31	5·28	5·27	5·25	5·24
4	4·54	4·32	4·19	4·11	4·05	4·01	3·98	3·95	3·94
5	4·06	3·78	3·62	3·52	3·45	3·40	3·37	3·34	3·32
6	3·78	3·46	3·29	3·18	3·11	3·05	3·01	2·98	2·96
7	3·59	3·26	3·07	2·96	2·88	2·83	2·78	2·75	2·72
8	3·46	3·11	2·92	2·81	2·73	2·67	2·62	2·59	2·56
9	3·36	3·01	2·81	2·69	2·61	2·55	2·51	2·47	2·44
10	3·29	2·92	2·73	2·61	2·52	2·46	2·41	2·38	2·35
11	3·23	2·86	2·66	2·54	2·45	2·39	2·34	2·30	2·27
12	3·18	2·81	2·61	2·48	2·39	2·33	2·28	2·24	2·21
13	3·14	2·76	2·56	2·43	2·35	2·28	2·23	2·20	2·16
14	3·10	2·73	2·52	2·39	2·31	2·24	2·19	2·15	2·12
15	3·07	2·70	2·49	2·36	2·27	2·21	2·16	2·12	2·09
16	3·05	2·67	2·46	2·33	2·24	2·18	2·13	2·09	2·06
17	3·03	2·64	2·44	2·31	2·22	2·15	2·10	2·06	2·03
18	3·01	2·62	2·42	2·29	2·20	2·13	2·08	2·04	2·00
19	2.99	2·61	2·40	2·27	2·18	2·11	2·06	2·02	1·98
20	2·97	2·59	2·38	2·25	2·16	2·09	2·04	2·00	1·96
21	2·96	2·57	2·36	2·23	2·14	2·08	2·02	1·98	1·95
22	2·95	2·56	2·35	2·22	2·13	2·06	2·01	1·97	1.93
23	2·94	2·55	2·34	2·21	2·11	2·05	1·99	1·95	1·92
24	2·93	2·54	2·33	2·19	2·10	2·04	1·98	1·94	1·91
25	2·92	2·53	2·32	2·18	2·09	2·02	1·97	1·93	1 89
26	2·91	2·52	2·31	2·17	2·08	2·01	1·96	1·92	1·88
27	2·90	2·51	2·30	2·17	2·07	2·00	1·95	1·91	1·87
28	2·89	2·50	2·29	2·16	2·06	2·00	1·94	1·90	1·87
29	2·89	2·50	2·28	2·15	2·06	1·99	1·93	1·89	1·86
30	2·88	2·49	2·28	2·14	2·05	1·98	1·93	1·88	1·85
40	2·84	2·44	2·23	2·09	2·00	1·93	1·87	1·83	1·79
60	2·79	2·39	2·18	2·04	1·95	1·87	1·82	1·77	1·74
120	2·75	2·35	2·13	1·99	1·90	1·82	1·77	1·72	1·68
∞	2·71	2·30	2·08	1·94	1·85	1·77	1·72	1·67	1·63

TABLE D7 (*continued*)

	m									
n	10	12	15	20	24	30	40	60	120	∞
1	60·19	60·71	61·22	61·74	62·00	62·26	62·53	62·79	63·06	63·33
2	9·39	9·41	9·42	9·44	9·45	9·46	9·47	9·47	9·48	9·49
3	5·23	5·22	5·20	5·18	5·18	5·17	5·16	5·15	5·14	5·13
4	3·92	3·90	3·87	3·84	3·83	3·82	3·80	3·79	3·78	3·76
5	3·30	3·27	3·24	3·21	3·19	3·17	3·16	3·14	3·12	3·10
6	2·94	2·90	2·87	2·84	2·82	2·80	2·78	2·76	2·74	2·72
7	2·70	2·67	2·63	2·59	2·58	2·56	2·54	2·51	2·49	2·47
8	2·54	2·50	2·46	2·42	2·40	2·38	2·36	2·34	2·32	2·29
9	2·42	2·38	2·34	2·30	2·28	2·25	2·23	2·21	2·18	2·16
10	2·32	2·28	2·24	2·20	2·18	2·16	2·13	2·11	2·08	2·06
11	2·25	2·21	2·17	2·12	2·10	2·08	2·05	2·03	2·00	1·97
12	2·19	2·15	2·10	2·06	2·04	2·01	1·99	1·96	1·93	1·90
13	2·14	2·10	2·05	2·01	1·98	1·96	1·93	1·90	1·88	1·85
14	2·10	2·05	2·01	1·96	1·94	1·91	1·89	1·86	1·83	1·80
15	2·06	2·02	1·97	1·92	1·90	1·87	1·85	1·82	1·79	1·76
16	2·03	1·99	1·94	1·89	1·87	1·84	1·81	1·78	1·75	1·72
17	2·00	1·96	1·91	1·86	1·84	1·81	1·78	1·75	1·72	1·69
18	1·98	1·93	1·89	1·84	1·81	1·78	1·75	1·72	1·69	1·66
19	1·96	1·91	1·86	1·81	1·79	1·76	1·73	1·70	1·67	1·63
20	1·94	1·89	1·84	1·79	1·77	1·74	1·71	1·68	1·64	1·61
21	1·92	1·87	1·83	1·78	1·75	1·72	1·69	1·66	1·62	1·59
22	1·90	1·86	1·81	1·76	1·73	1·70	1·67	1·64	1·60	1·57
23	1·89	1·84	1·80	1·74	1·72	1·69	1·66	1·62	1·59	1·55
24	1·88	1·83	1·78	1·73	1·70	1·67	1·64	1·61	1·57	1.53
25	1·87	1·82	1·77	1·72	1·69	1·66	1·63	1·59	1·56	1·52
26	1·86	1·81	1·76	1·71	1·68	1·65	1·61	1·58	1·54	1·50
27	1·85	1·80	1·75	1·70	1·67	1·64	1·60	1·57	1·53	1·49
28	1·84	1·79	1·74	1·69	1·66	1·63	1·59	1·56	1·52	1·48
29	1·83	1·78	1·73	1·68	1·65	1·62	1·58	1·55	1·51	1·47
30	1·82	1·77	1·72	1·67	1·64	1·61	1·57	1·54	1·50	1·46
40	1·76	1·71	1·66	1·61	1·57	1·54	1·51	1·47	1·42	1·38
60	1·71	1·66	1·60	1·54	1·51	1·48	1·44	1·40	1·35	1·29
120	1·65	1·60	1·55	1·48	1·45	1·41	1·37	1·32	1·26	1·19
∞	1·60	1·55	1·49	1·42	1·38	1·34	1·30	1·24	1·17	1·00

TABLE D7 (*continued*)

$F(F; m, n) = 0{\cdot}95$

	m								
n	1	2	3	4	5	6	7	8	9
1	161·4	199·5	215·7	224·6	230·2	234·0	236·8	238·9	240·5
2	18·51	19·00	19·16	19·25	19·30	19·33	19·35	19·37	19·38
3	10·13	9·55	9·28	9·12	9·01	8·94	8·89	8·85	8·81
4	7·71	6·94	6·59	6·39	6·26	6·16	6·09	6·04	6·00
5	6·61	5·79	5·41	5·19	5·05	4·95	4·88	4·82	4·77
6	5·99	5·14	4·76	4·53	4·39	4·28	4·21	4·15	4·10
7	5·59	4·74	4·35	4·12	3·97	3·87	3·79	3·73	3·68
8	5·32	4·46	4·07	3·84	3·69	3·58	3·50	3·44	3·39
9	5·12	4·26	3·86	3·63	3·48	3·37	3·29	3·23	3·18
10	4·96	4·10	3·71	3·48	3·33	3·22	3·14	3·07	3·02
11	4·84	3·98	3·59	3·36	3·20	3·09	3·01	2·95	2·90
12	4·75	3·89	3·49	3·26	3·11	3·00	2·91	2·85	2·80
13	4·67	3·81	3·41	3·18	3·03	2·92	2·83	2·77	2·71
14	4·60	3·74	3·34	3·11	2·96	2·85	2·76	2·70	2·65
15	4·54	3·68	3·29	3·06	2·90	2·79	2·71	2·64	2·59
16	4·49	3·63	3·21	3·01	2·85	2·74	2·66	2·59	2·54
17	4·45	3·59	3·20	2·96	2·81	2·70	2·61	2·55	2·49
18	4·41	3·55	3·16	2·93	2·77	2·66	2·58	2·51	2·46
19	4·38	3·52	3·13	2·90	2·74	2·63	2·54	2·48	2·42
20	4·35	3·49	3·10	2·87	2·71	2·60	2·51	2·45	2·39
21	4·32	3·47	3·07	2·84	2·68	2·57	2·49	2·42	2·37
22	4·30	3·44	3·05	2·82	2·66	2·55	2·46	2·40	2·34
23	4·28	3·42	3·03	2·80	2·64	2·53	2·44	2·37	2·32
24	4·26	3·40	3·01	2·78	2·62	2·51	2·42	2·36	2·30
25	4·24	3·39	2·99	2·76	2·60	2·49	2·40	2·34	2·28
26	4·23	3·37	2·98	2·74	2·59	2·47	2·39	2·32	2·27
27	4·21	3·35	2·96	2·73	2·57	2·46	2·37	2·31	2·25
28	4·20	3·34	2·95	2·71	2·56	2·45	2·36	2·29	2·24
29	4·18	3·33	2·93	2·70	2·55	2·43	2·35	2·28	2·22
30	4·17	3·32	2·92	2·69	2·53	2·42	2·23	2·27	2·21
40	4·08	3·23	2·84	2·61	2·45	2·34	2·25	2·18	2·12
60	4·00	3·15	2·76	2·53	2·37	2·25	2·17	2·10	2·04
120	3·92	3·07	2·68	2·45	2·29	2·17	2·09	2·02	1·96
∞	3·84	3·00	2·60	2·37	2·21	2·10	2·01	1·94	1·88

TABLE D7 (*continued*)

	m									
n	10	12	15	20	24	30	40	60	120	∞
1	241·9	243·9	245·9	248·0	249·1	250·1	251·1	252·2	253·3	254·3
2	19·40	19·41	19·43	19·45	19·45	19·46	19·47	19·48	19·49	19·50
3	8·79	8·74	8·70	8·66	8·64	8·62	8·59	8·57	8·55	8·53
4	5·96	5·91	5·86	5·80	5·77	5·75	5·72	5·69	5·66	5·63
5	4·74	4·68	4·62	4·56	4·53	4·50	4·46	4·43	4·40	4·36
6	4·06	4·00	3·94	3·87	3·84	3·81	3·77	3·74	3·70	3·67
7	3·64	3·57	3·51	3·44	3·41	3·38	3·34	3·30	3·27	3·23
8	3·35	3·28	3·22	3·15	3·12	3·08	3·04	3·01	2·97	2·93
9	3·14	3·07	3·01	2·94	2·90	2·86	2·83	2·70	2·75	2·71
10	2·98	2·91	2·85	2·77	2·74	2·70	2·66	2·62	2·58	2·54
11	2·85	2·79	2·72	2·65	2·61	2·57	2·53	2·49	2·45	2·40
12	2·75	2·69	2·62	2·54	2·51	2·47	2·43	2·38	2·34	2·30
13	2·67	2·60	2·53	2·46	2·42	2·38	2·34	2·30	2·25	2·21
14	2·60	2·53	2·46	2·39	2·35	2·31	2·27	2·22	2·18	2·13
15	2·54	2·48	2·40	2·33	2·29	2·25	2·20	2·16	2·11	2·07
16	2·49	2·42	2·35	2·28	2·24	2·19	2·15	2·11	2·06	2·01
17	2·45	2·38	2·31	2·23	2·19	2·15	2·10	2·06	2·01	1·96
18	2·41	2·34	2·27	2·19	2·15	2·11	2·06	2·02	1·97	1·92
19	2·38	2·31	2·23	2·16	2·11	2·07	2·03	1·98	1·93	1·88
20	2·35	2·28	2·20	2·12	2·08	2·04	1·99	1·95	1·90	1·84
21	2·32	2·25	2·18	2·10	2·05	2·01	1·96	1·92	1·87	1·81
22	2·30	2·23	2·15	2·07	2·03	1·98	1·94	1·89	1·84	1·78
23	2·27	2·20	2·13	2·05	2·01	1·96	1·91	1·86	1·81	1·76
24	2·25	2·18	2·11	2·03	1·98	1·94	1·89	1·84	1·79	1·73
25	2·24	2·16	2·09	2·01	1·96	1·92	1·87	1·82	1·77	1·71
26	2·22	2·15	2·07	1·99	1·95	1·90	1·85	1·80	1·75	1·69
27	2·20	2·13	2·06	1·97	1·93	1·88	1·84	1·79	1·73	1·67
28	2·19	2·12	2·04	1·96	1·91	1·87	1·82	1·77	1·71	1·65
29	2·18	2·10	2·03	1·94	1·90	1·85	1·81	1·75	1·70	1·64
30	2·16	2·09	2·01	1·93	1·89	1·84	1·79	1·74	1·68	1·62
40	2·08	2·00	1·92	1·84	1·79	1·74	1·69	1·64	1·58	1·51
60	1·99	1·92	1·84	1·75	1·70	1·65	1·59	1·53	1·47	1·39
120	1·91	1·83	1·75	1·66	1·61	1·55	1·50	1·43	1·35	1·25
∞	1·83	1·75	1·67	1·57	1·52	1·46	1·39	1·32	1·22	1·00

TABLE D7 (*continued*)

$F(F; m, n) = 0{\cdot}975$

	m								
n	1	2	3	4	5	6	7	8	9
1	647·8	799·5	864·2	899·6	921·8	937·1	948·2	956·7	963·3
2	38·51	39·00	39·17	39·25	39·30	39·33	39·36	39·37	39·39
3	17·44	16·04	15·44	15·10	14·88	14·73	14·62	14·54	14·47
4	12·22	10·65	9·98	9·60	9·36	9·20	9·07	8·98	8·90
5	10·01	8·43	7·76	7·39	7·15	6·98	6·85	6·76	6·68
6	8·81	7·26	6·60	6·23	5·99	5·82	5·70	5·60	5·52
7	8·07	6·54	5·89	5·52	5·29	5·12	4·99	4·90	4·82
8	7·57	6·06	5·42	5·05	4·82	4·65	4·53	4·43	4·36
9	7·21	5·71	5·08	4·72	4·48	4·32	4·20	4·10	4·03
10	6·94	5·46	4·83	4·47	4·24	4·07	3·95	3·85	3·78
11	6·72	5·29	4·63	4·28	4·04	3·88	3·76	3·66	3·59
12	6·55	5·10	4·47	4·12	3·89	3·73	3·61	3·51	3·44
13	6·41	4·97	4·35	4·00	3·77	3·60	3·48	3·39	3·31
14	6·30	4·86	4·24	3·89	3·66	3·50	3·38	3·29	3·21
15	6·20	4·77	4·15	3·80	3·58	3·41	3·29	3·20	3·12
16	6·12	4·69	4·08	3·73	3·50	3·34	3·22	3·12	3·05
17	6·04	4·62	4·01	3·66	3·44	3·28	3·16	3·06	2·98
18	5·98	4·56	3·95	3·61	3·38	3·22	3·10	3·01	2·93
19	5·92	4·51	3·90	3·56	3·33	3·17	3·05	2·96	2·88
20	5·87	4·46	3·86	3·51	3·29	3·13	3·01	2·91	2·84
21	5·83	4·42	3·82	3·48	3·25	3·09	2·97	2·87	2·80
22	5·79	4·38	3·78	3·44	3·22	3·05	2·93	2·84	2·76
23	5·75	4·35	3·75	3·41	3·18	3·02	2·90	2·81	2·73
24	5·72	4·32	3·72	3·38	3·15	2·99	2·87	2·78	2·70
25	5·69	4·29	2·69	3·35	3·13	2·97	2·85	2·75	2·68
26	5·66	4·27	3·67	3·33	3·10	2·94	2·82	2·73	2·65
27	5·63	4·24	3·65	3·31	3·08	2·92	2·80	2·71	2·63
28	5·61	4·22	3·63	3·29	3·06	2·90	2·78	2·69	2·61
29	5·59	4·20	3·61	3·27	3·04	2·88	2·76	2·67	2·59
30	5·57	4·18	3·59	3·25	3·03	2·87	2·75	2·65	2·57
40	5·42	4·05	3·46	3·13	2·90	2·74	2·62	2·53	2·45
60	5·29	3·93	3·34	3·01	2·79	2·63	2·51	2·41	2·33
120	5·15	3·80	3·23	2·89	2·67	2·52	2·39	2·30	2·22
∞	5·02	3·69	3·12	2·79	2·57	2·41	2·29	2·19	2·11

TABLE D7 (*continued*)

	m									
n	10	12	15	20	24	30	40	60	120	∞
1	968·6	976·7	984·9	993·1	997·2	1001	1006	1010	1014	1018
2	39·40	39·41	39·43	39·45	39·46	39·46	39·47	39·48	39·49	39·50
3	14·42	14·34	14·25	14·17	14·12	14·08	14·04	13·99	13·95	13·90
4	8·84	8·75	8·66	8·56	8·51	8·46	8·41	8·36	8·31	8·26
5	6·62	6·52	6·43	6·33	6·28	6·23	6·18	6·12	6·07	6·02
6	5·46	5·37	5·27	5·17	5·12	5·07	5·01	4·96	4·90	4·85
7	4·76	4·67	4·57	4·47	4·42	4·36	4·31	4·25	4·20	4·14
8	4·30	4·20	4·10	4·00	3·95	3·89	3·84	3·78	3·73	3·67
9	3·96	3·87	3·77	3·67	3·61	3·56	3·51	3·45	3·39	3·33
10	3·72	3·62	3·52	3·42	3·37	3·31	3·26	3·20	3·14	3·08
11	3·53	3·43	3·33	3·23	3·17	3·12	3·06	3·00	2·94	2·88
12	3·37	3·28	3·18	3·07	3·02	2·96	2·91	2·85	2·79	2·72
13	3·25	3·15	3·05	2·95	2·89	2·84	2·78	2·72	2·66	2·60
14	3·15	3·05	2·95	2·84	2·79	2·73	2·67	2·61	2·55	2·49
15	3·06	2·96	2·86	2·76	2·70	2·64	2·59	2·52	2·46	2·40
16	2·99	2·89	2·79	2·68	2·63	2·57	2·51	2·45	2·38	2·32
17	2·92	2·82	2·72	2·62	2·56	2·50	2·44	2·38	2·32	2·25
18	2·87	2·77	2·67	2·56	2·50	2·44	2·38	2·32	2·26	2·19
19	3·82	2·72	2·62	2·51	2·45	2·39	2·33	2·27	2·20	2·13
20	2·77	2·68	2·57	2·46	2·41	2·35	2·29	2·22	2·16	2·09
21	2·73	2·64	2·53	2·42	2·37	2·31	2·25	2·18	2·11	2·04
22	2·70	2·60	2·50	2·39	2·33	2·27	2·21	2·14	2·08	2·00
23	2·67	2·57	2·47	2·36	2·30	2·24	2·18	2·11	2·04	1·97
24	2·64	2·54	2·44	2·33	2·27	2·21	2·15	2·08	2·01	1·94
25	2·61	2·51	2·41	2·30	2·24	2·18	2·12	2·05	1·98	1·91
26	2·59	2·49	2·39	2·28	2·22	2·16	2·09	2·03	1·95	1·88
27	2·57	2·47	2·36	2·25	2·19	2·13	2·07	2·00	1·93	1·85
28	2·55	2·45	2·34	2·23	2·17	2·11	2·05	1·98	1·91	1·83
29	2·53	2·43	2·32	2·21	2·15	2·09	2·03	1·96	1·89	1·81
30	2·51	2·41	2·31	2·20	2·14	2·07	2·01	1·94	1·87	1·79
40	2·39	2·29	2·18	2·07	2·01	1·94	1·88	1·80	1·72	1·64
60	2·27	2·17	2·06	1·94	1·88	1·82	1·74	1·67	1·58	1·48
120	2·16	2·05	1·94	1·82	1·76	1·69	1·61	1·53	1·43	1·31
∞	2·05	1·94	1·83	1·71	1·64	1·57	1·48	1·39	1·27	1·00

TABLE D7 (*continued*)

$F(F; m, n) = 0{\cdot}99$

	m								
n	1	2	3	4	5	6	7	8	9
1	4052	4999·5	5403	5625	5764	5859	5928	5982	6022
2	98·50	99·00	99·17	99·25	99·30	99·33	99·36	99·37	99·39
3	34·12	30·82	29·46	28·71	28·24	27·91	27·67	27·49	27·35
4	21·20	18·00	16·69	15·98	15·52	15·21	14·98	14·80	14·66
5	16·26	13·27	12·06	11·39	10·97	10·67	10·46	10·29	10·16
6	13·75	10·92	9·78	9·15	8·75	8·47	8·26	8·10	7·98
7	12·25	9·55	8·45	7·85	7·46	7·19	6·99	6·84	6·72
8	11·26	8·65	7·59	7·01	6·63	6·37	6·18	6·03	5·91
9	10·56	8·02	6·99	6·42	6·06	5·80	5·61	5·47	5·35
10	10·04	7·56	6·55	5·90	5·64	5·39	5·20	5·06	4·94
11	9·65	7·21	6·22	5·67	5·32	5·07	4·89	4·74	4·63
12	9·33	6·93	5·95	5·41	5·06	4·82	4·64	4·50	4·39
13	9·07	6·70	5·74	5·21	4·86	4·62	4·44	4·30	4·19
14	8·86	6·51	5·56	5·04	4·69	4·46	4·28	4·14	4·03
15	8·68	6·36	5·42	4·89	4·56	4·32	4·14	4·00	3·89
16	8·53	6·23	5·29	4·77	4·44	4·20	4·03	3·89	3·78
17	8·40	6·11	5·18	4·67	4·34	4·10	3·93	3·79	3·68
18	8·29	6·01	5·09	4·58	4·25	4·01	3·84	3·71	3·60
19	8·18	5·93	5·01	4·50	4·17	3·94	3·77	3·63	3·52
20	8·10	5·85	4·94	4·43	4·10	3·87	3·70	3·56	3·46
21	8·02	5·78	4·87	4·37	4·04	3·81	3·64	3·51	3·40
22	7·95	5·72	4·82	4·31	3·99	3·76	3·59	3·45	3·35
23	7·88	5·66	4·76	4·26	3·94	3·71	3·54	3·41	3·30
24	7·82	5·61	4·72	4·22	3·90	3·67	3·50	3·36	3·26
25	7·77	5·57	4·68	4·18	3·85	3·63	3·46	3·32	3·22
26	7·72	5·53	4·64	4·14	3·82	3·59	3·42	3·29	3·18
27	7·68	5·49	4·60	4·11	3·78	3·56	3·39	3·26	3·15
28	7·64	5·45	4·57	4·07	3·75	3·53	3·36	3·23	3·12
29	7·60	5·42	4·54	4·04	3·73	3·50	3·33	3·20	3·09
30	7·56	5·39	4·51	4·02	3·70	3·47	3·30	3·17	3·07
40	7·31	5·18	4·31	3·83	3·51	3·29	3·12	2·99	2·89
60	7·08	4·98	4·13	3·65	3·34	3·12	2·95	2·82	2·72
120	6·85	4·79	3·95	3·48	3·17	2·96	2·79	2·66	2·56
∞	6·63	4·61	3·78	3·32	3·02	2·80	2·64	2·51	2·41

TABLE D7 (*continued*)

	m									
n	10	12	15	20	24	30	40	60	120	∞
1	6056	6106	6157	6209	6235	6261	6287	6313	6339	6366
2	99·40	99·42	99·43	99·45	99·46	99·47	99·47	99·48	99·49	99·50
3	27·23	27·05	26·87	26·69	26·60	26·50	26·41	26·32	26·22	26·13
4	14·55	14·37	14·20	14·02	13·93	13·84	13·75	13·65	13·56	13·46
5	10·05	9·89	9·72	9·55	9·47	9·38	9·29	9·20	9·11	9·02
6	7·87	7·72	7·56	7·40	7·31	7·23	7·14	7·06	6·97	6·88
7	6·62	6·47	6·31	6·16	6·07	5·99	5·91	5·82	5·74	5·65
8	5·81	5·67	5·52	5·36	5·28	5·20	5·12	5·03	4·95	4·86
9	5·26	5·11	4·96	4·81	4·73	4·56	4·57	4·48	4·40	4·31
10	4·85	4·71	4·56	4·41	4·33	4·25	4·17	4·08	4·00	3·91
11	4·54	4·40	4·25	4·10	4·02	3·94	3·86	3·78	3·69	3·60
12	4·30	4·16	4·01	3·86	3·78	3·70	3·62	3·54	3·45	3·36
13	4·10	3·96	3·82	3·66	3·59	3·51	3·43	3·34	3·25	3·17
14	3·94	3·80	3·66	3·51	3·43	3·35	3·27	3·18	3·09	3·00
15	3·80	3·67	3·52	3·37	3·29	3·21	3·13	3·05	2·96	2·87
16	3·69	3·55	3·41	3·26	3·18	3·10	3·02	2·93	2·84	2·75
17	3·59	3·46	3·31	3·16	3·08	3·00	2·92	2·83	2·75	2·65
18	3·51	3·37	3·23	3·08	3·00	2·92	2·84	2·75	2·66	2·57
19	3·43	3·30	3·15	3·00	2·92	2·84	2·76	2·67	2·58	2·49
20	3·37	3·23	3·09	2·94	2·86	2·78	2·69	2·61	2·52	2·42
21	3·31	3·17	3·03	2·88	2·80	2·72	2·64	2·55	2·46	2·36
22	3·26	3·12	2·98	2·83	2·75	2·67	2·58	2·50	2·40	2·31
23	3·21	3·07	2·93	2·78	2·70	2·62	2·54	2·45	2·35	2·26
24	3·17	3·03	2·89	2·74	2·65	2·58	2·49	2·40	2·31	2·21
25	3·13	2·99	2·85	2·70	2·62	2·54	2·45	2·36	2·27	2·17
26	3·09	2·96	2·81	2·66	2·58	2·50	2·42	2·33	2·23	2·13
27	3·06	2·93	2·78	2·63	2·55	2·47	2·38	2·29	2·20	2·10
28	3·03	2·90	2·75	2·60	2·52	2·44	2·35	2·26	2·17	2·06
29	3·00	2·87	2·73	2·57	2·49	2·41	2·33	2·23	2·14	2·03
30	2·98	2·84	2·70	2·55	2·47	2·39	2·30	2·21	2·11	2·01
40	2·80	2·66	2·52	2·37	2·29	2·20	2·11	2·02	1·92	1·80
60	2·63	2·50	2·35	2·20	2·12	2·03	1·94	1·84	1·73	1·60
120	2·47	2·34	2·19	2·03	1·95	1·86	1·76	1·66	1·53	1·38
∞	2·32	2·18	2·04	1·88	1·79	1·70	1·59	1·47	1·32	1·00

TABLE D7 (*continued*)

$F(F; m, n) = 0{\cdot}995$

	m								
n	1	2	3	4	5	6	7	8	9
1	16211	20000	21615	22500	23056	23437	23715	23925	24091
2	198·5	199·0	199·2	199·2	199·3	199·3	199·4	199·4	199·4
3	55·55	49·80	47·47	46·19	45·39	44·84	44·43	44·13	43·88
4	31·33	26·28	24·26	23·15	22·46	21·97	21·62	21·35	21·14
5	22·78	18·31	16·53	15·56	14·94	14·51	14·20	13·96	13·77
6	18·63	14·54	12·92	12·03	11·46	11·07	10·70	10·57	10·39
7	16·24	12·40	10·88	10·05	9·52	9·16	8·89	8·68	8·51
8	14·69	11·04	9·60	8·81	8·30	7·95	7·69	7·50	7·34
9	13·61	10·11	8·72	7·96	7·47	7·13	6·88	6·69	6·54
10	12·83	9·43	8·08	7·34	6·87	6·54	6·30	6·12	5·97
11	12·23	8·91	7·60	6·88	6·42	6·10	5·86	5·68	5·54
12	11·75	8·51	7·23	6·52	6·07	5·76	5·52	5·35	5·20
13	11·37	8·19	6·93	6·23	5·79	5·48	5·25	5·08	4·94
14	11·06	7·92	6·68	6·00	5·56	5·26	5·03	4·86	4·72
15	10·80	7·70	6·48	5·80	5·37	5·07	4·85	4·67	4·54
16	10·58	7·51	6·30	5·64	5·21	4·91	4·69	4·52	4·38
17	10·38	7·35	6·16	5·50	5·07	4·78	4·56	4·39	4·25
18	10·22	7·21	6·03	5·37	4·96	4·66	4·44	4·28	4·14
19	10·07	7·09	5·92	5·27	4·85	4·56	4·34	4·18	4·04
20	9·94	6·99	5·82	5·17	4·76	4·47	4·26	4·09	3·96
21	9·83	6·89	5·73	5·09	4·68	4·39	4·18	4·01	3·88
22	9·73	6·81	5·65	5·02	4·61	4·32	4·11	3·94	3·81
23	9·63	6·73	5·58	4·95	4·54	4·26	4·05	3·88	3·75
24	9·55	6·66	5·52	4·89	4·49	4·20	3·99	3·83	3·69
25	9·48	6·60	5·46	4·84	4·43	4·15	3·94	3·78	3·64
26	9·41	6·54	5·41	4·79	4·38	4·10	3·89	3·73	3·60
27	9·34	6·49	5·36	4·74	4·34	4·06	3·85	3·69	3·56
28	9·28	6·44	5·23	4·70	4·30	4·02	3·81	3·65	3·52
29	9·23	6·40	5·28	4·66	4·26	3·98	3·77	3·61	3·48
30	9·18	6·35	5·24	4·62	4·23	3·95	3·74	3·58	3·45
40	8·83	6·07	4·98	4·37	3·99	3·71	3·51	3·25	3·22
60	8·49	5·79	4·73	4·14	3·76	3·49	3·29	3·13	3·01
120	8·18	5·54	4·50	3·92	3·55	3·28	3·09	2·93	2·81
∞	7·88	5·30	4·28	3·72	3·35	3·09	2·90	2·74	2·62

TABLE D7 (*continued*)

	m									
n	10	12	15	20	24	30	40	60	120	∞
1	24224	24426	24630	24836	24940	25044	25148	25253	25359	25465
2	199·4	199·4	199·4	199·4	199·5	199·5	199·5	199·5	199·5	199·5
3	43·69	43·39	43·08	42·78	42·62	42·47	42·31	42·15	41·99	41·83
4	20·97	20·70	20·44	20·17	20·03	19·89	19·75	19·61	19·47	19·32
5	13·62	13·38	13·15	12·90	12·78	12·66	12·53	12·40	12·27	12·14
6	10·25	10·03	9·81	9·59	9·47	9·36	9·24	9·12	9·00	8·88
7	8·38	8·18	7·97	7·75	7·65	7·53	7·42	7·31	7·19	7·08
8	7·21	7·01	6·81	6·61	6·50	6·40	6·29	6·18	6·06	5·95
9	6·42	6·23	6·03	5·83	5·73	5·62	5·52	5·41	5·30	5·19
10	5·85	5·66	5·47	5·27	5·17	5·07	4·97	4·86	4·75	4·64
11	5·42	5·24	5·05	4·86	4·76	4·65	4·55	4·44	4·34	4·23
12	5·09	4·91	4·72	4·53	4·43	4·33	4·23	4·12	4·01	3·90
13	4·82	4·64	4·46	4·27	4·17	4·07	3·97	3·87	3·76	3·65
14	4·60	4·43	4·25	4·06	3·96	3·86	3·76	3·66	3·55	3·44
15	4·42	4·25	4·07	3·88	3·79	3·69	3·58	3·48	3·37	3·26
16	4·27	4·10	3·92	3·73	3·64	3·54	3·44	3·33	3·22	3·11
17	4·14	3·97	3·79	3·61	3·51	3·41	3·31	3·21	3·10	2·98
18	4·03	3·86	3·68	3·50	3·40	3·30	3·20	3·10	2·99	2·87
19	3·93	3·76	3·59	3·40	3·31	3·21	3·11	3·00	2·89	2·78
20	3·85	3·68	3·50	3·32	3·22	3·12	3·02	2·92	2·81	2·69
21	3·77	3·60	3·43	3·24	3·15	3·05	2·95	2·84	2·73	2·61
22	3·70	3·54	3·36	3·18	3·08	2·98	2·88	2·77	2·66	2·55
23	3·64	3·47	3·30	3·12	3·02	2·92	2·82	2·71	2·60	2·48
24	3·59	3·42	3·25	3·00	2·97	2·87	2·77	2·66	2·55	2·43
25	3·54	3·37	3·20	3·01	2·92	2·82	2·72	2·61	2·50	2·38
26	3·49	3·33	3·15	2·97	2·87	2·77	2·67	2·56	2·45	2·33
27	3·45	3·28	3·11	2·93	2·83	2·73	2·63	2·52	2·41	2·25
28	3·41	3·25	3·07	2·89	2·79	2·69	2·59	2·48	2·37	2·29
29	3·38	3·21	3·04	2·86	2·76	2·66	2·56	2·45	2·33	2·24
30	3·34	3·18	3·01	2·82	2·73	2·63	2·52	2·42	2·30	2·18
40	3·12	2·95	2·78	2·60	2·50	2·40	2·30	2·18	2·06	1·93
60	2·90	2·74	2·57	2·39	2·29	2·19	2·08	1·96	1·83	1·69
120	2·71	2·54	2·37	2·19	2·09	1·98	1·87	1·75	1·61	1·43
∞	2·52	2·36	2·19	2·00	1·90	1·79	1·67	1·53	1·36	1·00

TABLE D7 (*continued*)

$F(F; m, n) = 0{\cdot}999$

	m								
n	1	2	3	4	5	6	7	8	9
1	4053*	5000*	5404*	5625*	5764*	5859*	5929*	5981*	6023*
2	998·5	999·0	999·2	999·2	999·3	999·3	999·4	999·4	999·4
3	167·0	148·5	141·1	137·1	134·6	132·8	131·6	130·6	129·9
4	74·14	61·25	56·18	53·44	51·71	50·53	49·66	49·00	48·47
5	47·18	37·12	33·20	31·09	29·75	23·84	28·16	27·64	27·24
6	35·51	27·00	23·70	21·92	20·81	20·03	19·46	19·03	18·69
7	29·25	21·69	18·77	17·19	16·21	15·52	15·02	14·63	14·33
8	25·42	18·49	15·83	14·39	13·49	12·86	12·40	12·04	11·77
9	22·86	16·39	13·90	12·56	11·71	11·13	10·70	10·37	10·11
10	21·04	14·91	12·55	11·28	10·48	9·92	9·52	9·20	8·96
11	19·69	13·81	11·56	10·35	9·58	9·05	8·66	8·35	8·12
12	18·64	12·97	10·80	9·63	8·89	8·38	8·00	7·71	7·48
13	17·81	12·31	10·21	9·07	8·35	7·86	7·49	7·21	6·98
14	17·14	11·78	9·73	8·62	7·92	7·43	7·08	6·80	6·58
15	16·59	11·34	9·34	8·25	7·57	7·09	6·74	6·47	6·26
16	16·12	10·97	9·00	7·94	7·27	6·81	6·46	6·19	5·98
17	15·72	10·66	8·73	7·68	7·02	6·56	6·22	5·96	5·75
18	15·38	10·39	8·49	7·46	6·81	6·35	6·02	5·76	5·56
19	15·08	10·16	8·28	7·26	6·62	6·18	5·85	5·59	5·39
20	14·82	9·95	8·10	7·10	6·46	6·02	5·69	5·44	5·24
21	14·59	9·77	7·94	6·95	6·32	5·88	5·56	5·31	5·11
22	14·38	9·61	7·80	6·81	6·19	5·76	5·44	5·19	4·99
23	14·19	9·47	7·67	6·69	6·08	5·65	5·33	5·09	4 89
24	14·03	9·34	7·55	6·59	5·98	5·55	5·23	4·99	4·80
25	13·88	9·22	7·45	6·49	5·88	5·46	5·15	4·91	4·71
26	13·74	9·12	7·36	6·41	5·80	5·38	5·07	4·83	4·64
27	13·61	9·02	7·27	6·33	5·73	5·31	5·00	4·76	4·57
28	13·50	8·93	7·19	6·25	5·66	5·24	4·93	4·60	4·50
29	13·39	8·85	7·12	6·19	5·59	5·18	4·87	4·64	4·45
30	13·29	8·77	7·05	6·12	5·53	5·12	4·82	4·58	4·39
40	12·61	8·25	6·60	5·70	5·13	4·73	4·44	4·21	4·02
60	11·97	7·76	6·17	5·31	4·76	4·37	4·09	3·87	3·69
120	11·38	7·32	5·79	4·95	4·42	4·04	3·77	3·55	3·38
∞	10·83	6·91	5·42	4·62	4·10	3·74	3·47	3·27	3·10

* Multiply these entries by 100.

TABLE D7 (*continued*)

	m									
n	10	12	15	20	24	30	40	60	120	∞
1	6056*	6107*	6158*	6209*	6235*	6261*	6287*	6313*	6340*	6366*
2	999·4	999·4	999·4	999·4	999·5	999·5	999·5	999·5	999·5	999·5
3	129·2	128·3	127·4	126·4	125·9	125·4	125·0	124·5	124·0	123·5
4	48·05	47·41	46·76	46·10	45·77	45·43	45·09	44·75	44·40	44·05
5	26·92	26·42	25·91	25·39	25·14	24·87	24·60	24·33	24·06	23·79
6	18·41	17·99	17·56	17·12	16·89	16·67	16·44	16·21	15·99	15·75
7	14·08	13·71	13·32	12·93	12·73	12·53	12·33	12·12	11·91	11·70
8	11·54	11·19	10·84	10·48	10·30	10·11	9·92	9·73	9·53	9·33
9	9·89	9·57	9·24	8·90	8·72	8·55	8·37	8·19	8·00	7·81
10	8·75	8·45	8·13	7·80	7·64	7·47	7·30	7·12	6·94	6·76
11	7·92	7·63	7·32	7·01	6·85	6·68	6·52	6·35	6·17	6·00
12	7·29	7·00	6·71	6·40	6·25	6·09	5·93	5·76	5·59	5·42
13	6·80	6·52	6·23	5·93	5·78	5·63	5·47	5·30	5·14	4·97
14	6·40	6·13	5·85	5·56	5·41	5·25	5·10	4·94	4·77	4·60
15	6·08	5·81	5·54	5·25	5·10	4·95	4·80	4·64	4·47	4·31
16	5·81	5·55	5·27	4·99	4·85	4·70	4·54	4·39	4·23	4·06
17	5·58	5·32	5·05	4·78	4·63	4·48	4·33	4·18	4·02	3·85
18	5·39	5·13	4·87	4·59	4·45	4·30	4·15	4·00	3·84	3·67
19	5·22	4·97	4·70	4·43	4·20	4·14	3·99	3·84	3·68	3·51
20	5·08	4·82	4·56	4·29	4·15	4·00	3·86	3·70	3·54	3·38
21	4·95	4·70	4·44	4·17	4·03	3·88	3·74	3·58	3·42	3·26
22	4·83	4·58	4·33	4·06	3·92	3·78	3·63	3·48	3·32	3·15
23	4·73	4·48	4·23	3·96	3·82	3·68	3·53	3·38	3·22	3·05
24	4·64	4·39	4·14	3·87	3·74	3·59	3·45	3·29	3·14	2·97
25	4·56	4·31	4·06	3·79	3·66	3·52	3·37	3·22	3·06	2·89
26	4·48	4·24	3·99	3·72	3·59	3·44	3·30	3·15	2·99	2·82
27	4·41	4·17	3·92	3·66	3·52	3·38	3·23	3·08	2·92	2·75
28	4·35	4·11	3·86	3·60	3·46	3·32	3·18	3·02	2·86	2·69
29	4·29	4·05	3·80	3·54	3·41	3·27	3·12	2·97	2·81	2·64
30	4·24	4·00	3·75	3·49	3·36	3·22	3·07	2·92	2·76	2·59
40	3·87	3·64	3·40	3·15	3·01	2·87	2·73	2·57	2·41	2·23
60	3·54	3·31	3·08	2·83	2·69	2·55	2·41	2·25	2·08	1·89
120	3·24	3·02	2·78	2·53	2·40	2·26	2·11	1·95	1·76	1·54
∞	2·96	2·74	2·51	2·27	2·13	1·99	1·84	1·66	1·45	1·00

* Multiply these entries by 100.

Table D8

This table gives the sample size needed, for values of $P[\text{I}] = \alpha$ and $P[\text{II}] = 1 - \beta$, for a test on a single mean with unknown standard deviation. For example, if the null hypothesis is $H_0 : \mu = \mu_0$, and the alternative is $H_a : \mu < \mu_0$ the test statistic

$$t = (\bar{x} - \mu_0)\frac{\sqrt{n}}{s},$$

has a t-distribution with $(n - 1)$ degrees of freedom if H_0 is true, and the critical region $t > t_{1-\alpha}\ (n - 1)$ would have a significance level α. The table gives the sample size needed to control the values of α and β for various values of the auxiliary variable $\Delta = |\mu - \mu_0|/\sigma$, for both one-sided and two-sided tests.

	Level of t-test																				
Single-sided test Double-sided test	$\alpha = 0{\cdot}005$ $\alpha = 0{\cdot}01$					$\alpha = 0{\cdot}01$ $\alpha = 0{\cdot}02$					$\alpha = 0{\cdot}025$ $\alpha = 0{\cdot}05$					$\alpha = 0{\cdot}05$ $\alpha = 0{\cdot}1$					
$1 - \beta =$	0·01	0·05	0·1	0·2	0·5	0·01	0·05	0·1	0·2	0·5	0·01	0·05	0·1	0·2	0·5	0·01	0·05	0·1	0·2	0·5	
0·05																					0·05
0·10																					0·10
0·15																				122	0·15
0·20										139					99					70	0·20
0·25					110					90				128	64			139	101	45	0·25
0·30				134	78				115	63			119	90	45		122	97	71	32	0·30
0·35			125	99	58			109	85	47		109	88	67	34		90	72	52	24	0·35
0·40		115	97	77	45		101	85	66	37	117	84	68	51	26	101	70	55	40	19	0·40
0·45		92	77	62	37	110	81	68	53	30	93	67	54	41	21	80	55	44	33	15	0·45
0·50	100	75	63	51	30	90	66	55	43	25	76	54	44	34	18	65	45	36	27	13	0·50
0·55	83	63	53	42	26	75	55	46	36	21	63	45	37	28	15	54	38	30	22	11	0·55
0·60	71	53	45	36	22	63	47	39	31	18	53	38	32	24	13	46	32	26	19	9	0·60
0·65	61	46	39	31	20	55	41	34	27	16	46	33	27	21	12	39	28	22	17	8	0·65
0·70	53	40	34	28	17	47	35	30	24	14	40	29	24	19	10	34	24	19	15	8	0·70
0·75	47	36	30	25	16	42	31	27	21	13	35	26	21	16	9	30	21	17	13	7	0·75

Value of $\Delta = \frac{\mu - \mu_0}{\sigma}$																					
0·80	41	32	27	22	14	37	28	24	19	12	31	22	19	15	9	27	19	15	12	6	0·80
0·85	37	29	24	20	13	33	25	21	17	11	28	21	17	13	8	24	17	14	11	6	0·85
0·90	34	26	22	18	12	29	23	19	16	10	25	19	16	12	7	21	15	13	10	5	0·90
0·95	31	24	20	17	11	27	21	18	14	9	23	17	14	11	7	19	14	11	9	5	0·95
1·00	28	22	19	16	10	18	13	11	8	5	21	16	13	10	6	18	13	11	8	5	1·00
1·1	24	19	16	14	9	21	16	14	12	8	18	13	11	9	6	15	11	9	7		1·1
1·2	21	16	14	12	8	18	14	12	10	7	15	12	10	8	5	13	10	8	6		1·2
1·3	18	15	13	11	8	16	13	11	9	6	14	10	9	7		11	8	7	6		1·3
1·4	16	13	12	10	7	14	11	10	9	6	12	9	8	7		10	8	7	5		1·4
1·5	15	12	11	9	7	13	10	9	8	6	11	8	7	6		9	7	6			1·5
1·6	13	11	10	8	6	12	10	9	7	5	10	8	7	6		8	6	6			1·6
1·7	12	10	9	8	6	11	9	8	7		9	7	6	5		8	6	5			1·7
1·8	12	10	9	8	6	10	8	7	7		8	7	6			7	6				1·8
1·9	11	9	8	7	6	10	8	7	6		8	6	6			7	5				1·9
2·0	10	8	8	7	5	9	7	7	6		7	6	5			6					2·0
2·1	10	8	7	7		8	7	6	6		7	6				6					2·1
2·2	9	8	7	6		8	7	6	5		7	6				6					2·2
2·3	9	7	7	6		8	6	6			6	5				5					2·3
2·4	8	7	7	6		7	6	6			6										2·4
2·5	8	7	6	6		7	6	6			6										2·5
3·0	7	6	6	5		6	5	5			5										3·0
3·5	6	5	5			5															3·5
4·0	6																				4·0

TABLE D9

This table gives the sample size needed, for given values of $P[\mathrm{I}] = \alpha$ and $P[\mathrm{II}] = 1 - \beta$, for a test of the hypothesis of the equality of two means, $H_0: \mu_1 = \mu_2$, where there is a common, but unknown variance. The test statistic is

$$t = \frac{\bar{x}_1 - \bar{x}_2}{s[(1/n_1 + 1/n_2)]^{\frac{1}{2}}},$$

where

$$s = \frac{(n_1 - 1)\, s_1{}^2 + (n_2 - 1)\, s_2{}^2}{n_1 + n_2 - 2},$$

which is distributed as student's t-distribution with $(n_1 + n_2 - 2)$ degrees of freedom if H_0 is true. This table is used to obtain sample sizes needed to control the values of α and β for various values of the auxiliary parameter $\Delta = (\mu_1 - \mu_2)/\sigma$, for both one-sided and two-sided tests.

	Level of t-test																				
Single-sided test Double-sided test	$\alpha = 0{\cdot}005$ $\alpha = 0{\cdot}01$					$\alpha = 0{\cdot}01$ $\alpha = 0{\cdot}02$					$\alpha = 0{\cdot}025$ $\alpha = 0{\cdot}05$					$\alpha = 0{\cdot}05$ $\alpha = 0{\cdot}1$					
$1 - \beta =$	0·01	0·05	0·1	0·2	0·5	0·01	0·05	0·1	0·2	0·5	0·01	0·05	0·1	0·2	0·5	0·01	0·05	0·1	0·2	0·5	
0·05																					0·05
0·10																					0·10
0·15																					0·15
0·20																				137	0·20
0·25															124					88	0·25
0·30										123					87					61	0·30
0·35					110					90					64				102	45	0·35
0·40					85					70				100	50			108	78	35	0·40
0·45				118	68				101	55			105	79	39		108	86	62	28	0·45
0·50				96	55			106	82	45		106	86	64	32		88	70	51	23	0·50

Value of $\Delta = \frac{\mu_1 - \mu_2}{\sigma}$																					
0·55			101	79	46		106	88	68	38		87	71	53	27	112	73	58	42	19	0·55
0·60		101	85	67	39		90	74	58	32	104	74	60	45	23	89	61	49	36	16	0·60
0·65		87	73	57	34	104	77	64	49	27	88	63	51	39	20	76	52	42	30	14	0·65
0·70	100	75	63	50	29	90	66	55	43	24	76	55	44	34	17	66	45	36	26	12	0·70
0·75	88	66	55	44	26	79	58	48	38	21	67	48	39	29	15	57	40	32	23	11	0·75
0·80	77	58	49	39	23	70	51	43	33	19	59	42	34	26	14	50	35	28	21	10	0·80
0·85	69	51	43	35	21	62	46	38	30	17	52	37	31	23	12	45	31	25	18	9	0·85
0·90	62	46	39	31	19	55	41	34	27	15	47	34	27	21	11	40	28	22	16	8	0·90
0·95	55	42	35	28	17	50	37	31	24	14	42	30	25	19	10	36	25	20	15	7	0·95
1·00	50	38	32	26	15	45	33	28	22	13	38	27	23	17	9	33	23	18	14	7	1·00
1·1	42	32	27	22	13	38	28	23	19	11	32	23	19	14	8	27	19	15	12	6	1·1
1·2	36	27	23	18	11	32	24	20	16	9	27	20	16	12	7	23	16	13	10	5	1·2
1·3	31	23	20	16	10	28	21	17	14	8	23	17	14	11	6	20	14	11	9	5	1·3
1·4	27	20	17	14	9	24	18	15	12	8	20	15	12	10	6	17	12	10	8	4	1·4
1·5	24	18	15	13	8	21	16	14	11	7	18	13	11	9	5	15	11	9	7	4	1·5
1·6	21	16	14	11	7	19	14	12	10	6	16	12	10	8	5	14	10	8	6	4	1·6
1·7	19	15	13	10	7	17	13	11	9	6	14	11	9	7	4	12	9	7	6	3	1·7
1·8	17	13	11	10	6	15	12	10	8	5	13	10	8	6	4	11	8	7	5		1·8
1·9	16	12	11	9	6	14	11	9	8	5	12	9	7	6	4	10	7	6	5		1·9
2·0	14	11	10	8	6	13	10	9	7	5	11	8	7	6	4	9	7	6	4		2·0
2·1	13	10	9	8	5	12	9	8	7	5	10	8	6	5	3	8	6	5	4		2·1
2·2	12	10	8	7	5	11	9	7	6	4	9	7	6	5		8	6	5	4		2·2
2·3	11	9	8	7	5	10	8	7	6	4	9	7	6	5		7	5	5	4		2·3
2·4	11	9	8	6	5	10	8	7	6	4	8	6	5	4		7	5	4	4		2·4
2·5	10	8	7	6	4	9	7	6	5	4	8	6	5	4		6	5	4	3		2·5
3·0	8	6	6	5	4	7	6	5	4	3	6	5	4	4		5	4	3			3·0
3·5	6	5	5	4	3	6	5	4	4		5	4	4	3		4	3				3·5
4·0	6	5	4	4		5	4	4	3		4	4	3			4					4·0

TABLE D10

This table gives the value of the ratio R of the population variance σ^2 to a standard variance $\sigma_0{}^2$, which is undetected with probability $\bar{\beta} = 1 - \beta$ in a χ^2 test at significance level α of an estimate s^2 of σ^2 based on n degrees of freedom. For $R < 1$ enter the table with $\bar{\beta}' = \alpha$, $\alpha' = \bar{\beta}$ and $R' = 1/R$.

	$\alpha = 0{\cdot}01$				$\alpha = 0{\cdot}05$			
n	$\bar{\beta} = 0{\cdot}01$	$\bar{\beta} = 0{\cdot}05$	$\bar{\beta} = 0{\cdot}1$	$\bar{\beta} = 0{\cdot}5$	$\bar{\beta} = 0{\cdot}01$	$\bar{\beta} = 0{\cdot}05$	$\bar{\beta} = 0{\cdot}1$	$\bar{\beta} = 0{\cdot}5$
1	42,240	1,687	420·2	14·58	25,450	977·0	243·3	8·444
2	458·2	89·78	43·71	6·644	298·1	58·40	28·43	4·322
3	98·79	32·24	19·41	4·795	68·05	22·21	13·37	3·303
4	44·69	18·68	12·48	3·955	31·93	13·35	8·920	2·826
5	27·22	13·17	9·369	3·467	19·97	9·665	6·875	2·544
6	19·28	10·28	7·628	3·144	14·44	7·699	5·713	2·354
7	14·91	8·524	6·521	2·911	11·35	6·491	4·965	2·217
8	12·20	7·352	5·757	2·736	9·418	5·675	4·444	2·112
9	10·38	6·516	5·198	2·597	8·103	5·088	4·059	2·028
10	9·072	5·890	4·770	2·484	7·156	4·646	3·763	1·960
12	7·343	5·017	4·159	2·312	5·889	4·023	3·335	1·854
15	5·847	4·211	3·578	2·132	4·780	3·442	2·925	1·743
20	4·548	3·462	3·019	1·943	3·802	2·895	2·524	1·624
24	3·959	3·104	2·745	1·842	3·354	2·630	2·326	1·560
30	3·403	2·752	2·471	1·735	2·927	2·367	2·125	1·492
40	2·874	2·403	2·192	1·619	2·516	2·103	1·919	1·418
60	2·358	2·046	1·902	1·490	2·110	1·831	1·702	1·333
120	1·829	1·661	1·580	1·332	1·686	1·532	1·457	1·228
∞	1·000	1·000	1·000	1·000	1·000	1·000	1·000	1·000

TABLE D11

This table shows the ratio R of two population variances, i.e. $R = \sigma_2^2/\sigma_1^2$, which remains undetected with probability $\bar{\beta} = 1 - \beta$ in a variance ratio test at significance level α of the ratio s_2^2/s_1^2 of estimates of the two variances, each being based on n degrees of freedom.

	α=0·01				α=0·05				α=0·5			
n	$\bar{\beta}$=0·01	$\bar{\beta}$=0·05	$\bar{\beta}$=0·1	$\bar{\beta}$=0·5	$\bar{\beta}$=0·01	$\bar{\beta}$=0·05	$\bar{\beta}$=0·1	$\bar{\beta}$=0·5	$\bar{\beta}$=0·01	$\bar{\beta}$=0·05	$\bar{\beta}$=0·1	$\bar{\beta}$=0·5
1	16,420,000	654,200	161,500	4052	654,200	26,070	6,436	161.5	4,052	161.5	39.85	1·000
2	9,000	1,881	891·0	99·00	1,881	361·0	171·0	19·00	99·00	19·00	9·000	1·000
3	867·7	273·3	158·8	29·46	273·3	86·06	50·01	9·277	29·46	9·277	5·391	1·000
4	255·3	102·1	65·62	15·98	102·1	40·81	26·24	6·388	15·98	6·388	4·108	1·000
5	120·3	55·39	37·87	10·97	55·39	25·51	17·44	5·050	10·97	5·050	3·453	1·000
6	71·67	36·27	25·86	8·466	36·27	18·35	13·09	4·284	8·466	4·284	3·056	1·000
7	48·90	26·48	19·47	6·993	26·48	14·34	10·55	3·787	6·993	3·787	2·786	1·000
8	36·35	20·73	15·61	6·029	20·73	11·82	8·902	3·438	6·029	3·438	2·589	1·000
9	28·63	17·01	13·06	5·351	17·01	10·11	7·757	3·179	5·351	3·179	2·440	1·000
10	23·51	14·44	11·26	4·849	14·44	8·870	6·917	2·978	4·849	2·978	2·323	1·000
12	17·27	11·16	8·923	4·155	11·16	7·218	5·769	2·687	4·155	2·687	2·147	1·000
15	12·41	8·466	6·946	3·522	8·466	5·777	4·740	2·404	3·522	2·404	1·972	1·000
20	8·630	6·240	5·270	2·938	6·240	4·512	3·810	2·124	2·938	2·124	1·794	1·000
24	7·071	5·275	4·526	2·659	5·275	3·935	3·376	1·984	2·659	1·984	1·702	1·000
30	5·693	4·392	3·833	2·386	4·392	3·389	2·957	1·841	2·386	1·841	1·606	1·000
40	4·470	3·579	3·183	2·114	3·579	2·866	2·549	1·693	2·114	1·693	1·506	1·000
60	3·372	2·817	2·562	1·836	2·817	2·354	2·141	1·534	1·836	1·534	1·396	1·000
120	2·350	2·072	1·939	1·533	2·072	1·828	1·710	1·352	1·533	1·352	1·265	1·000
∞	1·000	1·000	1·000	1·000	1·000	1·000	1·000	1·000	1·000	1·000	1·000	1·000

Bibliography

There exist numerous books on statistics and related subjects, and no attempt has been made to compile a comprehensive bibliography. Rather, the list below contains some of the classic books on the subject, plus others which I have found particularly readable and useful.

A. Probability Theory

H. Cramèr *The Elements of Probability Theory*, John Wiley (1955). An extremely readable introduction to the frequency theory of probability and probability distributions.

H. Cramèr *Random Variables and Probability Distributions*, Cambridge, (1962). A short, but more advanced, discussion of probability distributions.

W. Feller *An Introduction to Probability Theory and its Applications*, John Wiley (1957). A rigorous, but readable, book on most aspects of probability theory.

B. General Statistics

(i) *Introductory and Intermediate*

K. A. Brownlee *Statistical Theory and Methodology in Science and Engineering*, John Wiley (1960).

S. Ehrenfeld and S. B. Littauer *Introduction to Statistical Method*, McGraw-Hill (1964).

Guttman and Wilks *Introductory Engineering Statistics*, John Wiley.

A. M. Mood and F. A. Graybill *Introduction to the Theory of Statistics*, McGraw-Hill (1963).

All four of the above books provide well-written introductions to statistics, and would be suitable for a first reading in the subject.

B. W. Lindgren *Statistical Theory*, Macmillan (1968). An interesting book on statistics written at the intermediate level, but from the viewpoint of decision theory.

(ii) *Advanced*

T. W. Anderson *An Introduction to Multivariate Statistical Analysis*, John Wiley (1958). A very detailed book covering all aspects of multivariate analysis, least squares and hypothesis testing.

H. Cramèr *Mathematical Methods of Statistics*. Princeton (1946). A classic text on statistical theory which, like Cramèr's other works, is always very readable.

M. Kendall and A. Stuart *The Advanced Theory of Statistics*, Griffen (1958). An immense three-volume work, the first two volumes of which contain a wealth of detail about topics that we have discussed. Numerous examples, both worked and as exercises.

S. S. Wilks *Mathematical Statistics*, John Wiley (1962). A general advanced level text on theoretical statistics.

C. Estimation and Hypothesis Testing

P. R. Bevington *Data Reduction and Error Analysis for the Physical Sciences*, McGraw-Hill (1969). An elementary book on the treatment of errors, curve fitting etc., with the additional interesting feature of linking the text by a number of computer programs for statistical analysis.

F. A. Graybill *An Introduction to Linear Statistical Models*, McGraw-Hill (1961). A complete detailed treatment of regression and related topics.

W. C. Hamilton *Statistics in Physical Science*, Ronald (1964). An intermediate level book written by a chemist for other scientists. Good section on least-squares.

R. L. Plackett *Principles of Regression Analysis*, Oxford (1960). A very useful text on least-squares, regression, linear hypotheses, etc.

E. Lehmann *Testing Statistical Hypotheses*, John Wiley (1952). A complete text at an advanced and rigorous level.

D. Optimization Theory

M. J. Box, D. Davies and W. H. Swann *Non-linear Optimization Techniques*, Oliver and Boyd (1969). A short introduction to current techniques.

R. Fletcher (ed.) *Optimization*, Academic Press (1969). The proceedings of a Conference on Optimization held in 1968. Contains the latest views on the subject, plus useful reviews of recent progress.

J. Kowalik and M. R. Osborne *Methods for Unconstrained Optimization Problems*, American Elsevier (1968). A somewhat more detailed book that that of Box *et al.* which, despite the title, also treats constrained problems. Contains the results of a collection of test problems treated by several current methods.

E. Tables

W. H. Beyer (ed) *Handbook of Tables for Probability and Statistics*, Chemical Rubber Co. (1966). A large reference work.

E. S. Pearson and H. O. Hartley *Biometrika Tables for Statisticians, Vol. I.* Cambridge (1958). A short collection of the most useful statistical tables. Suitable for personal purchase.

Index

D

E

F

G

H

I

J

K

L

M